CONSUMER GUIDE
TO
HOME ENERGY SAVINGS

The **American Council for an Energy-Efficient Economy** (ACEEE) is a nonprofit organization dedicated to advancing energy efficiency as a means of promoting both economic prosperity and environmental protection. Support for our work comes from a wide range of foundations, government organizations, research institutes, utilities, and corporations.

ACEEE publishes guides for consumers and equipment procurement officials: *Guide to Energy-Efficient Office Equipment; Green Guide to Cars and Trucks: Model Years 1998 and 1999*; and *Guide to Energy-Efficient Commercial Equipment.*

Other titles include ACEEE's Book Series on Energy Policy: *Improving Energy Efficiency in Apartment Buildings; Energy-Efficient Motor Systems: A Handbook on Technology, Program, and Policy Opportunities; Transportation, Energy, and Environment: How Far Can Technology Take Us?; Transportation and Energy: Strategies for a Sustainable Transportation System; Transportation and Global Climate Change; Energy Efficiency and the Environment: Forging the Link; Residential Indoor Air Quality and Energy Efficiency;* and *Electric Utility Planning and Regulation.*

To request the ACEEE Publications Catalog, write ACEEE, 1001 Connecticut Avenue, N.W., Suite 801, Washington, D.C. 20036. Phone: (202) 429-0063. Fax: (202) 429-0193. World Wide Web: http://aceee.org.

Home Energy is a nonprofit magazine about residential energy conservation. Articles cover everything from research into new energy-saving technologies to tips on how to save energy in the home. Written for the professional and the homeowner, it is "the best magazine in America on home energy," according to John Javna, author of *50 Simple Things You can Do to Save the Earth.* To subscribe, write Home Energy, 2124 Kittredge Street, Suite 95, Berkeley, California 94704.

CONSUMER GUIDE TO HOME ENERGY SAVINGS

by
Alex Wilson, Jennifer Thorne, and John Morrill

Seventh Edition

American Council for an Energy-Efficient Economy
Washington, D.C.

IN COOPERATION WITH:
Home Energy magazine
Berkeley, California

Published by the American Council for an Energy-Efficient Economy, 1001 Connecticut Avenue, N.W., Suite 801, Washington, D.C. 20036.

Cover and book design by Chuck Myers
Illustrations by David Conover
Printed in the United States of America

Portions of this book have been reprinted with the permission of Massachusetts Audubon Society from the following guides: *Saving Energy and Money with Home Appliances; Oil and Gas Heating Systems; How to Weatherize Your Home or Apartment; All About Insulation*; and *Contractor's Guide to Sealing Air Leaks*.

Listings of the most efficient appliances are necessarily incomplete. Smaller manufacturers often do not have efficiency data independently monitored. Also, new, more efficient models are constantly joining the market. Please use this book as a guide, not a bible.

ISBN 0-918249-38-4

 Printed on Recycled Paper

TABLE OF CONTENTS

LISTS OF TABLES AND MOST EFFICIENT APPLIANCES

MOST EFFICIENT APPLIANCES

ACKNOWLEGMENTS

This book owes much to the energy conservation community: those hardworking folks who have worked diligently (and quietly) to increase the energy efficiency of the U.S. housing stock by one-third since 1973, and whose trials and errors have produced the knowledge compiled here.

The information in this guide has evolved with the steady improvements in residential appliances and changes in equipment efficiency standards. We thank Aimee Marciniak for her helpful research assistance with this seventh edition. Each new edition also benefits from the suggestions of experts in this field. We thank all the individuals who have reviewed this or previous editions for clarity and technical accuracy: Carl Blumstein, Fred Davis, Neal Elliott, Howard Geller, Roger Harris, John Hayes, Drew Kleibrink, Michael L'Ecuyer, Marc Ledbetter, Karina Lutz, Chris Mathis, Alan Meier, Steve Nadel, Nancy Schalch, Mike Thompson, and Hofu Wu.

Thanks also to David Conover for illustrating this guide in a clear and engaging manner, to Renee Nida, who skillfully copyedited this work, and to Glee Murray and Eric Stragar for their help with production. As usual, the authors remain responsible for any errors or omissions that remain.

CHAPTER 1
Saving Energy and Saving the Environment

A re you about to buy a new appliance? Remodel your house? Upgrade your heating or cooling system? If you're like most of us, you don't do these things very often. When you do, you want to make good choices, both for your pocketbook and for the environment. But you probably don't have time to become an expert. That's where this book can help.

The *Consumer Guide to Home Energy Savings* will help you decide which products to buy and how to use them for maximum energy savings. We've sorted through the thousands of major home appliances and heating systems on the market and picked out those that are the most efficient. We've listed the best ways to tighten up your house so that your heating and cooling systems won't have to work as hard—or use as much energy. We've pulled together tips on operating new and existing appliances to reduce energy use and improve performance. But before getting into the details, let's take a look at why it makes sense to buy the most efficient appliances and conserve energy in the home.

ENERGY USE AND THE ENVIRONMENT

Every time you buy a home appliance, tune up your heating system, or replace a burned-out light bulb, you're making a decision that affects the environment. You are probably already aware that most of our biggest environmental problems are directly associated with energy production and use: urban smog, oil spills, acid rain, and global warming, to mention a few. But you may not realize just how big a difference each of us can make by taking energy use into account in our household purchasing and maintenance decisions.

For example, did you know that every kilowatt-hour (kWh) of electricity you avoid using saves over two pounds of carbon dioxide that would otherwise be pumped into the atmosphere?[1] Carbon dioxide (CO_2) is the number one contributor to global warming, a process that scientists say could raise the earth's temperatures by 5–9°F over the

---

1 This assumes that your utility company produces electricity by burning coal. If another fuel or hydropower is used, less CO_2 will be produced.

next hundred years. If you replace a typical 1978, 18-cubic-foot refrigerator with an energy-efficient 1999 model, you'll save about 1,000 kWh and a ton of CO_2 emissions per year! Installing an 18-watt compact fluorescent light in place of a 75-watt incandescent light bulb will save about 570 kWh and over 1,300 pounds of CO_2 emissions over the life of the compact fluorescent.

Carbon dioxide is only one of the environmentally harmful gases resulting from energy use. Others, such as sulfur dioxide, nitrous oxide, carbon monoxide, and ozone, have much more direct effects—effects that can be seen and smelled in every major urban area of the country. That new refrigerator mentioned above will save each year over 20 pounds of sulfur dioxide emissions, the leading cause of acid rain.

Other energy-saving products and improvements around the home can help the environment. Table 1.1. shows the reductions in CO_2 emissions achieved from a few relatively easy energy improvements in the home. With some of these you'll notice different CO_2 savings depending on the type of fuel used. That's because some fuels give off less CO_2 than others. If you're interested in this kind of analysis, turn to Appendix 1 for details on the CO_2 emissions from different energy sources.

Worldwide, we pump some 20 billion tons of CO2 into the atmosphere each year—four tons for every man, woman, and child on earth. The United States is responsible for one-quarter of that, or 5 billion tons per year. On a per-capita basis, that comes to 18 tons for each American, though some of us produce a lot more than others.

Reducing CO_2 emissions by a few tons per year may not seem like a lot, given the billions of tons released worldwide each year. But the collective actions of many will have a dramatic effect, particularly in a high energy-use country like the United States. For example, the new appliance efficiency standards that went into effect at the beginning of 1990 will save more than 45 billion kWh in 2000—about 34 million tons of CO_2. If the roughly 40 million households in climates with large heating needs boosted their furnace or boiler efficiencies from 70% to 90%, some 45 million tons of CO_2 emissions would be eliminated each year. Substituting compact fluorescent lamps for the ten most frequently used incandescent lamps in every house in the country would reduce CO_2 emissions by about the same amount.

The U.S. Environmental Protection Agency (EPA) and U.S. Department of Energy (DOE) recognize the local and global environmental significance of residential appliances and equipment. Working in voluntary cooperation with manufacturers and retailers, these agencies have created a distinctive ENERGY STAR® label to help consumers identify energy-efficient appliances, computers, lighting, and home

TABLE 1.1

ENERGY CONSERVATION AND CO_2 SAVINGS IN THE HOME

Energy conservation measure	CO_2 Gas	savings Oil	(tons/yr) Electric[1]
Installing 10 13-watt compact fluorescent light bulbs in place of 10 60-watt incandescent bulbs[2]	—	—	1.1
Replacing typical 1978 refrigerator with energy-efficient 1998 model[3]	—	—	1.3
Replacing a 65% efficient furnace or boiler with one that is 90% efficient[4]	2.0	3.0	—
Substituting gas or oil heat for electric resistance heat[1,4]	23	19	—
Replacing single-glazed windows with argon-filled double-glazed windows[4]	2.4	3.9	9.8
Planting shade trees around house and painting house a lighter color[5]	—	—	.9–2.4
Installing a solar water-heating system[6]	.84	1.4	4.9
Boosting energy efficiency of house being built from standard insulation levels to super-insulated standards[7]	5.5	8.8	23

1. Assumes electricity generated using coal.
2. Assumes lights on 2,000 hours per year (5½ hours per day).
3. Average 1978 model uses 1,600 kWh per year; energy-efficient 1998 model uses 550 kWh/year.
4. Assumes 1,850 square-foot house of average (good) energy efficiency (heating load of 6.95 Btu/ft^2/°F-day) in a northern climate (6,300 heating degree-days).
5. Data from Lawrence Berkeley Laboratory, Berkeley, Calif. Based on computer simulations for various locations around the country.
6. Assumes two-panel system providing 14.25 million Btu/year (75% of demand).
7. Assumes 1,850 square-foot house in northern climate (6,300 degree-days). Boosting energy efficiency from 6.95 Btu/ft^2/°F-day to 1.37 Btu/ft^2/°F-day (going from R-19 walls, R-30 ceilings, double-glazed windows, and relatively loose construction to R-31 walls, R-38 ceilings, low-e windows and tight construction).

3

entertainment equipment. Some homebuilders are now constructing entire ENERGY STAR® homes, which include a variety of energy-efficient features and equipment. ENERGY STAR® homes are at least 30 percent more energy efficient than the current International Energy Conservation Code (formerly known as the Model Energy Code).

EPA   DOE

SAVING THE EARTH. SAVING YOUR MONEY.

To get a sense of just how effective energy conservation can be, take a look at the 1970s and 1980s. From 1973 to 1986, the U.S. gross national product grew 36% with no increase in energy use at all. Had efficiencies remained at 1973 levels, we would be spending an extra $150 billion in energy bills each year and pumping 1½ times more CO_2 into the atmosphere! We are already saving 13 million barrels of oil each day—half of the OPEC output—and, compared with 1973 projections, we're getting by with 250 fewer large power plants than would have otherwise been required.

The ENERGY STAR® logo identifies energy-efficient appliances and equipment. The ENERGY STAR® distinction is now available for qualifying furnaces, central and room air conditioners, heat pumps, refrigerators, dishwashers, clothes washers, light fixtures, televisions and VCRs, audio equipment, computers and computer monitors, and a variety of other office equipment.

DEPENDENCE ON FOREIGN OIL

The risks associated with heavy reliance on foreign oil were brought back into focus by the Iraqi invasion of Kuwait in August 1990 and the costly war that followed. More recently, the terrorist bombing against Americans stationed in Saudi Arabia underscored our vulnerability to that unstable region.

Lest we forget: in the late 1970s, the United States imported over 40% of its oil, with a peak of 46.4% in 1977. Then dramatic price increases and supply shortages fueled highly effective energy conservation efforts. Net petroleum imports dropped to 28% by 1982. But by the mid-80s, as fuel prices softened (due in part to successful conservation efforts), interest in conservation dropped and oil imports crept back up—to over 40%

since 1989. During several months of 1990, in fact, dependence on foreign oil reached an all-time high of over 50%.

The answer isn't to open up wilderness areas to oil exploration, but to use less energy more efficiently. A 1½ miles-per-gallon (mpg) increase in fuel economy by automobiles in the United States would save more oil than is estimated to lie under the Arctic National Wildlife Refuge. Just a ⅒ mpg increase would save more oil than is beneath the Georges Bank, a treasured fishing area off the Atlantic coast threatened by oil exploration. Adding low-emissivity coatings to all windows in the country would save an equivalent of ½ million barrels of oil per day—⅓ of the oil we import from Persian Gulf countries.

Indeed, energy efficiency improvements and aggressive energy conservation measures could totally eliminate U.S. dependency on foreign oil. We can't look just to the governing bodies to bring about these changes—it's up to each of us. And it should start in our homes and our cars.

SAVING THE EARTH—
AND GETTING PAID TO DO IT

The wonderful thing about saving energy is that, in addition to helping the environment, you save money. It's like contributing to a good cause and ending up with more money in your pocket. Many of the energy-efficient appliances and heating or cooling systems covered in this book cost no more than their inefficient counterparts. With most others, the extra cost is easily repaid in energy savings over just a few years.

When you buy an appliance, you pay more than just the sales price—you commit yourself to paying the cost of running the appliance for as long as you own it. These energy costs can add up quickly. For example, running a refrigerator 15-20 years costs two to three times as much as the initial

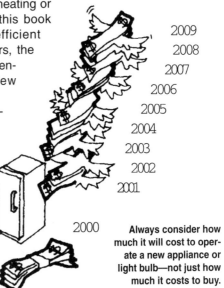

2009
2008
2007
2006
2005
2004
2003
2002
2001
2000

Always consider how much it will cost to operate a new appliance or light bulb—not just how much it costs to buy.

5

purchase price of the unit. That 100-watt light bulb you just put in will cost about $6 in electricity over its (short) life.

The sum of the purchase price and the energy cost of running an appliance or light bulb over its lifetime is called its life-cycle cost. The life-cycle costs of energy-efficient appliances are lower than those of average models even though the average-efficiency models may cost less to buy.

To increase the economic benefits of buying more energy-efficient appliances and boosting your overall home efficiency, your utility company may offer a rebate program for the purchase of energy-efficient appliances. Utility companies have realized that it costs less money to conserve energy than to build new power plants and that they can save money by paying you to save energy. Rebates are most common for high-efficiency heatpumps, central air conditioners, refrigerators, and clothes washers. Rebate programs are much more common among electric companies than gas companies, although some gas utilities offer rebates for high-efficiency furnaces and boilers. If you plan to buy a major appliance soon, ask your utility if it offers rebates for efficient models.

HOW TO USE THIS BOOK

The *Consumer Guide to Home Energy Savings* is the most complete and up-to-date guide available on energy savings in the home. Following a review of measures to tighten up the building shell itself, the book focuses on the things you put in it—including major appliances, heating equipment, air conditioning, and lighting—and how the energy use of those products can be reduced.

If you're about to buy a new appliance or heating system, you'll be most interested in the tips on what to look for when buying new equipment and the listings throughout this book of the most energy-efficient models available. Only the very highest rated models are listed within each appliance category. The models listed here represent only a small percentage of all the different models currently available. Many more appliances are better than average in terms of efficiency, but only the best are included here.

Our lists are based primarily on information from the most recent product directories published by appliance industry associations. We have also tried to include highly efficient models not listed in these directories. The efficiency ratings are based on standardized tests that manufacturers are required to conduct on their products. The same test ratings are listed on the yellow EnergyGuide labels that are required for most home appliances. See Appendix 2 for information on how to read and interpret EnergyGuide labels, and read through the detailed information on energy performance in each chapter.

The lists of refrigerators, freezers, water heaters, clothes washers, and dishwashers include the estimated annual energy use. The lists of furnaces, boilers, air conditioners, and heat pumps include

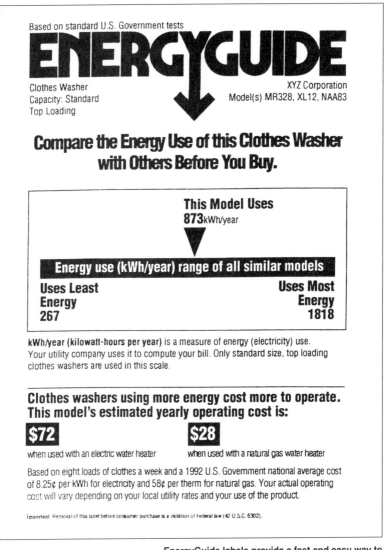

EnergyGuide labels provide a fast and easy way to compare the energy efficiency of most home appliances.

their efficiency ratings instead of estimated energy use. The energy use for these appliances varies greatly from house to house depending on climate, family size, and other factors, but the efficiency ratings provides a good way to compare the energy use of one model to another. The higher the efficiency, the lower the energy use and operating cost.

Keep in mind that energy performance is not the only consideration you should use when selecting home appliances. Consumers must consider how effectively they perform their primary functions—cleaning dishes, keeping food cold, etc.—as well as how much energy they use in doing it. For example, you wouldn't consider buying a dishwasher that didn't get your dishes clean, even if it used just half as much energy. This book does not pretend to be a comprehensive review of product reliability or performance, or a guide to convenience features found in these products; there are other sources for that information. It is worth noting, however, that energy-efficient appliances are generally high-quality products due to the better materials and components used in their construction.

The model numbers in this book are used by manufacturers in their product literature. In some cases, appliance dealers use abbreviated model numbers (for example, dropping the first number or letter). An asterisk (*) appearing in a model number indicates a digit or letter that varies with features of the appliance not affecting its energy efficiency and capacity (for example, color).

When shopping for major home appliances, you may want to call several stores or dealers to check the price and availability of different models. Ask the salesperson for information about the efficiency of each model, but be aware that he or she may not know very much about energy performance. Take this guide along when you shop. Use it along with the yellow EnergyGuide labels to compare similar models.

Several new appliance makers have entered the U.S. market in recent years. These and some other companies manufacturing high-efficiency equipment are not exactly "household names," and it may not be easy to find their products in your local appliance store. Appendix 3 lists phone numbers and, where available, web sites for all the companies whose products are listed in this guide. Customer service representatives should be able to direct you to a local distributor.

If you cannot find some of the models listed in this book, you can still use the information provided here to your advantage. Compare the efficiencies of the models you can find to those that are listed here. Don't be satisfied if the salesperson tells you that one model is just as efficient as another unless he or she can demonstrate this to you. Indeed, some brand-new models may be as efficient as those in this book, but were introduced after this book went to press. Use the models listed here for comparison.

And thank you for taking the time to consider energy efficiency. Remember, every time you do something to save energy, you're helping the environment. All of us—and especially our children and their children—owe you a big thanks.

CHAPTER 2
Buttoning Up Your House

If you live in a cold climate, you probably spend something like two-thirds of your energy dollars on heat. Your old furnace or boiler chugs away burning the gas or oil like there's no tomorrow. So should you rush right out and buy a new super-efficient one? Not necessarily.

Replacing your existing heating system with one that's more efficient may well be a wise step, but it shouldn't necessarily be your first step. You should first try to lower your heating requirements. Tighten up. Weatherize. Insulate. By reducing your heating needs, you may be able to get by with a significantly smaller—and less expensive—furnace or boiler.

The same arguments hold true with air conditioning. If you live in a warm climate with high cooling requirements, it makes a lot of sense to tighten up the house to reduce your cooling load before investing in new air conditioning equipment.

A tight, well-insulated house not only saves energy and allows you to get by with smaller-capacity heating and air conditioning systems, but it is also more comfortable. No more cold drafts at your feet while temperatures at head level are a sweaty 80°F. With less of this temperature stratification during the winter months, you'll even find yourself comfortable at a lower thermostat setting than you're used to.

In the next few pages, we'll take a look at some of the measures you can take to improve the energy efficiency of your house and when it makes sense to consider such projects.

CONSIDER AN ENERGY AUDIT

Find out from a pro where heat is being lost through your home's shell and what you should do about it. Energy auditors use sophisticated equipment, like a blower door and infrared camera, to help pinpoint air leaks and areas with inadequate insulation. Trained energy auditors know what to look for in both newer and older houses, and the investment in their time is usually well worth the cost. Depending on the service, you may be able to have your heating or cooling system cleaned, tuned up, and tested at the same time (see Chapters 4 and 5). Check with your utility company to find out if it offers this service. Some utility companies provide basic energy audits free of charge.

Some energy auditors and building scientists provide a wide range of services to improve the overall performance and comfort provided by houses. Sometimes called *house doctors* or *home performance contractors*, they address each building as a system. They understand how to improve a home's overall safety, comfort, energy efficiency, and indoor air quality. After all, many construction flaws can result in high energy bills, and some conditions that cause high energy bills can compromise building safety or resident health and comfort. For example, home performance contractors can diagnose and fix uneven heat distribution while lowering energy costs, or reduce cold drafts and moisture condensation problems while reducing energy bills. For more information on services by home performance contractors, visit this site on the World Wide Web: http://www.home-performance.org. Or contact your state energy office or cooperative extension service for information on qualified auditors and contractors.

Have an energy audit done on your house to find out where the heat really goes—and what you should do about it.

If you are a renter, encourage your landlord to have an energy audit conducted and follow through on the recommended energy improvements. You might offer to help your landlord by arranging for the audit and even doing some of the work in exchange for rent. After all, if you pay for heating, air conditioning, and electricity, energy improvements are very much in your interest. Even if your landlord won't pay for energy conservation projects, many of the suggestions in this chapter are inexpensive enough that they'll pay for themselves in just a year or two, justifying your out-of-pocket expenditures.

FIND AND SEAL AIR LEAKS

Hidden air leaks are among the largest heat loss sources in most older homes. Some of the most common air leakage sites are listed below:

- Plumbing penetrations through insulated floors and ceilings
- Chimney penetrations through insulated ceilings and exterior walls
- Along the sill plate and band joist at the top of foundation walls
- Fireplace dampers
- Attic access hatches
- The tops of interior partition walls where they intersect with the attic space
- Recessed lights and fans in insulated ceilings
- Wiring penetrations through insulated floors, ceilings, and walls
- Missing plaster
- Electrical outlets and switches, especially on exterior walls
- Window, door, and baseboard moldings
- Dropped ceilings above bathtubs and cabinets
- Kneewalls in finished attics, especially at access doors and built-in cabinets and bureaus.

What you should use in sealing these hidden air leaks depends on the size of the gaps and where they are located. Caulk is best for cracks and gaps less than about ¼" wide. In choosing caulks, read the label carefully to make sure that the caulk is suitable for the material to be sealed. Look for caulks that remain flexible over a 20-year lifetime. If the caulked joint will be visible, choose a paintable caulk or one that is the right color. In general, you should avoid the cheapest caulks, because they probably won't hold up well.

Expanding foam sealant is an excellent material to use for sealing larger cracks and holes that are protected from sunlight and moisture. One-part polyurethane foam is commonly available in hardware and building supply stores. Look for foam sealant without ozone-depleting CFCs (chlorofluorocarbons). Only buy products that are labeled as safe for the ozone.

Backer rod or crack filler is a flexible foam material, usually round in cross-section (¼" to 1" in diameter), and sold in long coils. Use it for sealing large cracks and to provide a backing in very deep cracks that are to be sealed with caulk.

Use rigid foam insulation for sealing very large openings such as plumbing chases and attic hatch covers. Fiberglass insulation can also be used for sealing large holes, but it will work better if wrapped in plastic or stuffed in plastic bags, because air can leak through exposed fiberglass. Don't use plastic in places where high temperatures may be

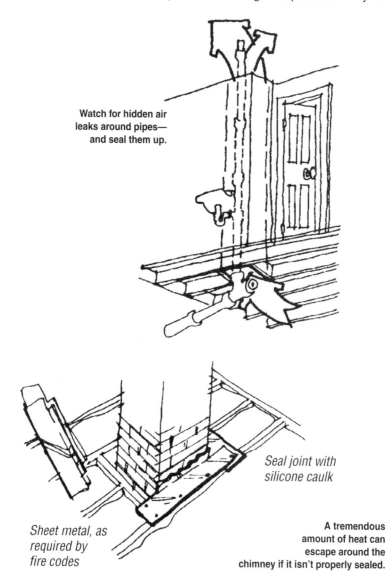

Watch for hidden air leaks around pipes— and seal them up.

Seal joint with silicone caulk

Sheet metal, as required by fire codes

A tremendous amount of heat can escape around the chimney if it isn't properly sealed.

reached, and always wear gloves and a dust mask when working with fiberglass.

Sheets of polyethylene can be taped over large holes to block air flow in some situations, but this is usually a fairly temporary measure, since the polyethylene may disintegrate over time if not protected.

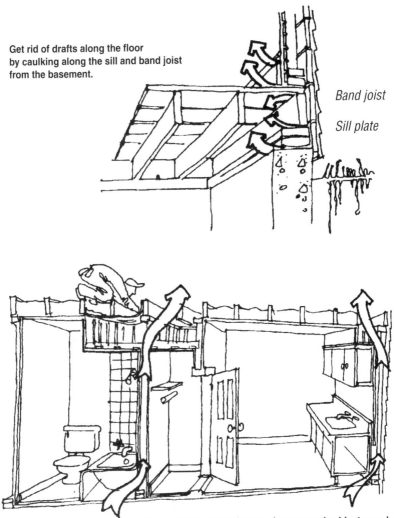

Get rid of drafts along the floor by caulking along the sill and band joist from the basement.

Band joist

Sill plate

Drop-ceilings above closets, showers, and cabinets can be among the worst offenders when it comes to air leakage.

Specialized materials such as metal flashing and high-temperature silicone sealants may be required for sealing around chimneys and flue pipes. Check with your building inspector or fire marshal if unsure about fire-safe details in these locations.

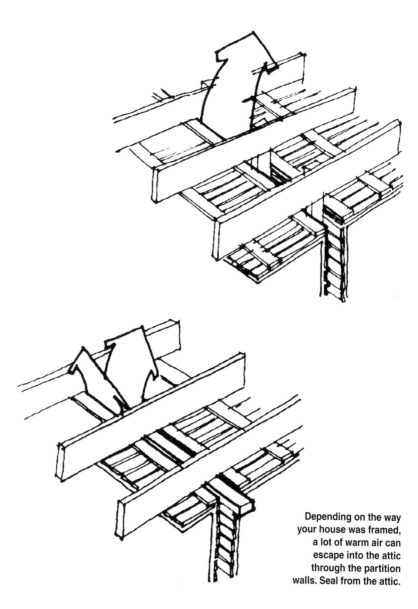

Depending on the way your house was framed, a lot of warm air can escape into the attic through the partition walls. Seal from the attic.

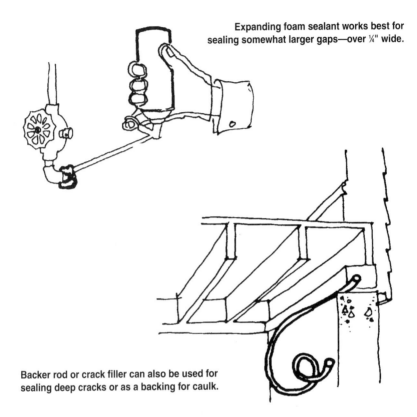

Expanding foam sealant works best for sealing somewhat larger gaps—over ¼" wide.

Backer rod or crack filler can also be used for sealing deep cracks or as a backing for caulk.

UPGRADE INEFFICIENT WINDOWS AND DOORS

About one-third of the home's total heat loss usually occurs through windows and doors. In fact, the total energy lost through windows and doors in this country is about as much as the amount of energy we get from the Alaska Pipeline! Windows deserve the most attention because they outnumber doors. To reduce heat loss, you can either fix up your existing windows or replace them with new energy-efficient units. Which choice you should take depends on a number of factors.

If the existing windows have rotted or damaged wood, cracked glass, missing putty, poorly fitting sashes, or locks that don't work, you may be better off replacing them. The next chapter discusses the exciting advances in new window technology. But if the windows are generally in good shape, it will probably be more cost-effective to boost their efficiency by weatherstripping, caulking, and fitting them with storm panels.

17

Outside

Inside

A tremendous amount of cold air can leak in through old windows.

New replacement windows will typically cost from $200 to $400 apiece, including labor. If you are going to the expense of installing new windows, be sure to spend the few extra dollars necessary to buy high-performance units with low-e (low emissivity) glass and argon gas fill between two or more panes of glass. The next chapter describes features of new windows in greater detail.

If your existing windows are in relatively good shape, it may be hard to justify the expense of window replacement. In that case, there are a number of ways to improve their energy efficiency. The quickest and least expensive option is to weatherstrip all window edges and cracks with rope caulk. This costs less than $1 per window and only takes a few minutes. Rope caulk may be taken off, stored in foil, and reused for two or three seasons, but once it hardens you should discard it.

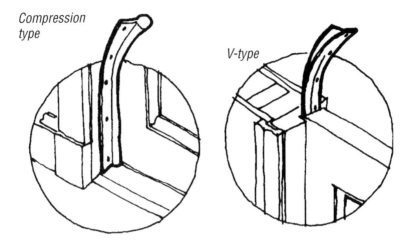

Compression type

V-type

Weatherstripping is the most permanent way to cut air leakage through windows and doors.

A more permanent solution is to weatherstrip the windows. This is more time-consuming and expensive than installing rope caulk ($8–$10 per window), but it only needs to be done once, it permits you to open the window, and the weatherstripping is out of sight. The type of weatherstripping to use depends on the type of window—both compression-type and V-strip weatherstripping are widely available in building supply stores. You usually get what you pay for with weatherstripping products so spend the few extra dollars necessary to get a top-quality product that is likely to hold up over years of use. With double-hung windows, if you don't need to open the upper sash, you can permanently caulk them closed.

The next step in improving window energy efficiency is to install some type of storm window. If you have single-glazed windows, storm panels will double their energy efficiency. If you heat with oil, the improved energy efficiency in a cold climate will save about a gallon of oil per square-foot of window per year. (That's an awful lot of oil that is otherwise leaking out of your house and adding to our air pollution woes.) With gas heat, the savings can top 1 therm per square foot per year.

The simplest type of storm window is a plastic film taped to the inside of the window frame. Inner storm window kits are readily available from hardware stores. They cost just $3–$8 per window and typically last for one to three years. Some are made of special shrink-tight plastic that you heat with a blow dryer after installation to get rid of wrinkles. These inexpensive plastic films are especially suitable for apartments and condominiums where exterior improvements are not allowed or not practical.

Removable or operable storm windows with glass or rigid acrylic panes generally make more sense if you plan to stay in the house for more than a few years. Both exterior and interior storm windows are available, though exterior units are far more common. Most people choose aluminum-framed combination storm/screen windows, which are very convenient to operate. Be careful, though. There are big differences among products on the market, especially relative to air tightness. The tightest units have air leakage rates as low as 0.01 cfm/ft, while the worst are over 1 cfm/ft. Don't settle for anything higher than 0.3 cfm/ft. Also look for units with low-e coatings on the glass to improve the energy performance.

Storm windows typically cost between $50 and $120, depending on size, quality, and labor for installation—far less than replacement windows. Before buying new storm windows, though, check to make sure

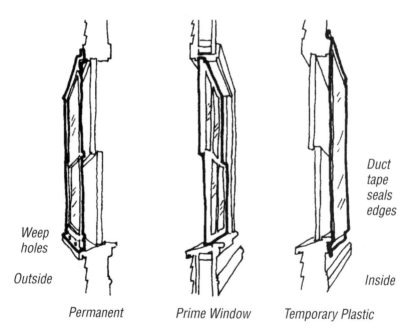

Weep holes

Outside

Duct tape seals edges

Inside

Permanent   *Prime Window*   *Temporary Plastic*

Storm windows can be installed on the inside or outside.

you don't already have a stack in a dusty corner of your basement or attic. If these are older wood-framed storm windows, they probably just need a coat of paint. Although they aren't quite as convenient as combination storm/screen windows (they have to be taken down and put up each year and separate screens are required), wooden storm windows are often more energy efficient.

If aluminum combination storms are already in place, examine them to make sure they are tightly sealed where mounted to the window casings; if not, caulk all cracks (but do not seal the small weep holes on the bottom edges).

Install door sweeps to reduce air flow underneath doors.

Finally, you can boost the energy efficiency of windows by installing insulating curtains or drapes on the interior. These can be closed at night to significantly cut down on heat loss. They can also be closed on hot summer days to keep out unwanted heat gain. Look for shades or drapes that fit into tracks to keep air from passing around the edges.

Don't forget about your doors. As with windows, make sure your doors are in good shape. Weatherstrip around the whole perimeter to ensure a tight seal when closed. Install quality door sweeps on the bottom of the doors if they aren't already in place. On an old, uninsulated metal or fiberglass door, a storm door probably isn't cost-effective. In fact, a glass storm door may damage the plastic trim on some metal or fiberglass doors by trapping heat.

INSULATE

Insulation is your primary defense against heat loss through the house envelope. However, putting insulation into a house after it is built can be pretty difficult. Because of the large area involved, walls are most important. You can probably tell if the walls are insulated by removing an outlet cover and peering into the wall cavity. Another way to check for insulation is to find a closet (or cabinet) along an exterior wall. Drill two ¼" holes into the wall about four inches apart, with one hole above the other. Shine a flashlight into one hole while looking into the other, and any insulation should be apparent. If there isn't any insulation, the best option is to bring in an insulation contractor to blow cellulose or fiberglass into the walls.

Adding insulation to an unheated attic is usually a lot easier. If there is no floor in the attic, simply add more insulation, either loose fill or unfaced fiberglass batts. If the existing insulation comes up to the top of the joists, add an additional layer of unfaced batts across the joists. This helps to cover gaps between the first layer of batts. If an attic floor is in place, you may need to remove that floor before adding insulation (be very careful not to step through the ceiling below!). In most of the country, a full foot of fiberglass or cellulose insulation is cost-effective in the attic floor.

If the attic is finished with a sloped cathedral ceiling, adding extra insulation is much more difficult. If there is no insulation there at present, it may be worth pulling off the drywall, insulating, and installing a new ceiling. Or, if you will be re-roofing in the near future, consider adding a

Cellulose or fiberglass insulation can be blown into uninsulated walls by an insulation contractor.

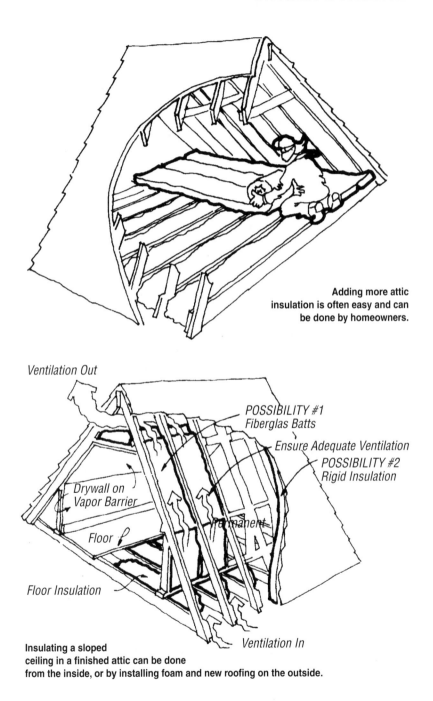

Adding more attic insulation is often easy and can be done by homeowners.

Ventilation Out

POSSIBILITY #1
Fiberglas Batts

Ensure Adequate Ventilation

POSSIBILITY #2
Rigid Insulation

Drywall on
Vapor Barrier

Permanent

Floor

Floor Insulation

Ventilation In

Insulating a sloped ceiling in a finished attic can be done from the inside, or by installing foam and new roofing on the outside.

23

layer of rigid foam insulation and decking on top of the existing roof and then shingling over that. In either case, it's a major project, and you will probably have to bring in a contractor to do the work. If there is already some insulation in the roof, it may be hard to justify the cost of this work, in which case you can focus your energy efforts elsewhere.

Heat loss through foundation walls is often neglected even in new houses. But in fact, in an otherwise well-insulated and tight house, as much as 20% of the total heat loss can occur through uninsulated

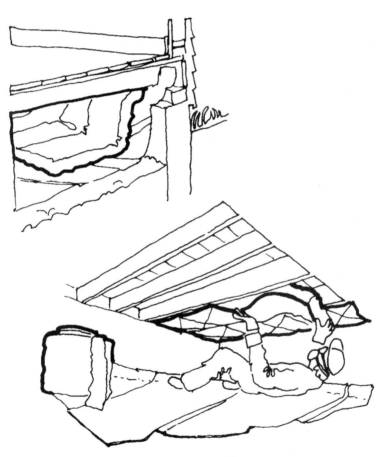

To reduce heat loss to an unheated
crawl space or basement, insulate between your floor joists.
If pipes in the crawl space could freeze, insulate around
the perimeter, as shown in the top illustration.

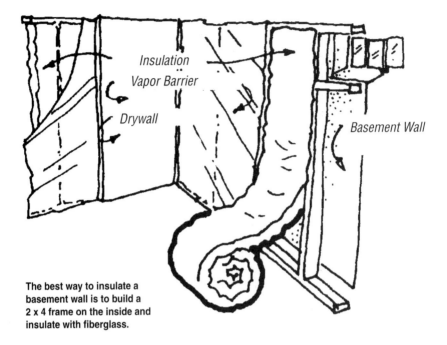

Insulation
Vapor Barrier
Drywall
Basement Wall

The best way to insulate a basement wall is to build a 2 x 4 frame on the inside and insulate with fiberglass.

foundation walls. Insulating the foundation or floor can easily save several hundred gallons of oil or several hundred therms of gas per year in northern climates.

If the basement or crawl space is unheated and you don't have plans to fix it up, you will do best to insulate between the floor joists instead of around the perimeter walls. Use unfaced fiberglass batt insulation supported from below with wire or metal rods if necessary. Fluff the insulation to ensure that it fills the joist cavities. Cover the underside of the joists with a moisture-permeable air barrier, such as Tyvek® or Typar®. In crawl spaces, cover the ground surface with 6-mil-thick polyethylene to keep moisture from getting into the crawl space from the ground.

If the basement is heated and used, you need to insulate the basement walls instead. The simplest method is to build 2 × 4 frames against the concrete foundation walls, insulate with fiberglass, and cover with drywall. Before doing this, you may need to correct drainage problems on the exterior if water leaks into the basement. Hire a reputable contractor who understands foundation drainage, or, if you do the work yourself, consult a good do-it-yourself manual.

ECONOMICS

The economics of all these energy improvements will depend on where you live, how large your heating or cooling requirements are, and how much you pay for energy. Most measures described in this chapter will pay for themselves in five years or less if you heat with gas or oil. If you have electric resistance heat, the payback will generally be much faster. Do-it-yourself energy improvements pay for themselves the fastest because the labor cost is eliminated—but make sure you know what you're doing!

PLANNING ENERGY IMPROVEMENTS

Some of the more involved energy improvements mentioned here, such as window replacement and insulating, make the most sense when you are planning other remodeling work. If you are going to extend a wall out to enlarge your kitchen or put in a larger dormer for a master bedroom expansion, by all means boost energy efficiency at the same time. Rebuild walls with high insulation levels. Put in high-performance insulating windows.

As long as you're ripping out walls, take advantage of the mess and go a little further, boosting the efficiency of some of the adjoining walls and windows as well. With a small addition, some of this work might even pay for itself right away if it means, for example, that you can get by without adding a separate heating system or expanding your current heat distribution system.

NEW CONSTRUCTION

If you are building a new house or a major addition, you can do it right the first time, saving money and the environment for decades to come. Today's state-of-the-art energy-efficient houses typically require less than a quarter as much energy for heating and cooling as most existing houses. There are thousands of homes in the northern United States and Canada with yearly energy bills that total just $200 to $300.

These homes cost more to build than a standard house, but not that much more. You might spend an extra $5,000–$10,000 to build a super-efficient house with R-30 walls, R-38 ceilings, R-19 foundations, R-3 windows, and very low air leakage. But that extra cost will usually be recovered in just five to ten years through energy savings. Plus, you'll be more comfortable and you'll have the satisfaction of knowing that your house is dumping less pollution and carbon dioxide into the atmosphere.

Once you get an idea of what you want, contact builders or architects in your area and find out how experienced they are with energy-efficient construction. Special skills are required to build high-efficiency houses and to install features such as heat-recovery ventilation systems. You may need to spend a little extra time looking for the right builder, but the time and effort will be well worth it.

CHAPTER 3
New Windows

Whether replacing windows in an older house or choosing windows for a new house, your decisions on what type of windows to buy will be among the most important decisions you will make in terms of energy use. In older houses, it is not uncommon for windows to account for a third of the total heat loss from a house. In new houses, windows typically account for 15% to 40% of total heat loss. Also, those windows can account for as much as 75% of heat gain during the summer months, adding to air conditioning costs. On the positive side, windows can be used for passive solar heating in the winter months and help reduce heating costs, and of course windows provide natural daylighting and views to the outside world.

Because of the impact windows have on both heat loss and heat gain, proper selection of products can be confusing. To add to the complexity, window glazing technology has changed tremendously in recent years. The best window glazings today insulate almost four times as well as the best commonly available windows just ten years ago. Because of the rapid pace of change, even skilled designers are often not fully aware of the potential these new glazings offer for energy-efficient building design.

UNDERSTANDING HOW WINDOWS WORK

To understand how windows affect heating and cooling costs, we need to know a little about how energy flows through them. The next illustration shows the primary mechanisms of heat transfer through windows.

■ *Sunlight* (solar radiation) is an important source of heat, and is transmitted directly through most windows. Solar radiation consists of visible light and a part of the solar spectrum that is heat but not visible light (infrared heat radiation). Glazings are available that *selectively* transmit light of different wavelengths—for example, some appear totally clear (high visible light transmittance) while blocking infrared heat radiation. A window's solar heat gain coefficient (SHGC) is the measure of the amount of solar energy that passes through the window; typical values range from 0.4 to 0.9, and the higher the

SHGC, the greater the percentage of solar energy that is transmitted to the inside.

■ *Radiant heat* is given off by warmer objects to colder objects. Things warmed by sunlight become stronger sources of radiant heat, and radiant heat is blocked by most window glazings. Greenhouses and passive solar heating illustrate the dynamic between sunlight and radiant heat: sunlight passes through glazings, warming objects indoors, but the heat from those objects does not quickly escape back through the glazings, and the space warms up. Special coatings on high-performance windows change the way heat radiation is absorbed and reradiated (see discussion on low-e glazings below).

■ *Conduction* is the mechanism of heat transfer through physical contact. Heat conducts from the warmer to the cooler side of a window as each molecule excites its neighbor, passing the energy along. Conduction occurs not only through solid materials (window glass and frames), but also through the air space between the layers of glass.

The amount of heat transmission through a material due to a temperature difference is given by its "U-value." The smaller the U-value, the less heat is transmitted. One of the major recent advances in window performance has been to use special gases between the layers of glass. These gases conduct heat less readily than air, resulting in lower U-values than simple double-glazed windows. **The NFRC uses the term "U-factor" in place of "U-value." The meaning is the same, and readers should consider the terms interchangeable.**

■ *Convection* is the movement of heat in a fluid, such as air. The heat is transferred as the molecules of air are physically moved from one place to another. A warm glass surface heats the air next to it, causing the air to rise. A cold glass surface is warmed by the air next to it, and that air mass will fall as it gives up its heat. These convection currents occur on the inside of a window, on the outside, and between layers of glass.

■ *Infiltration* is the process that carries heat through cracks and gaps around window frames. Infiltration through leaky windows can carry cold air into a house and warm air out. Infiltration is driven by wind and other differences in air pressure, such as warm air rising inside a house.

FEATURES TO LOOK FOR
IN ENERGY-SAVING WINDOWS

As mentioned above, window technology has improved dramatically in recent years, with the net result of lowering your energy bills.

Some of the most important energy features of windows are explained below.

■ *Multiple layers of glazing.* Until the 1980s the primary way manufacturers improved the energy performance of windows was to add additional layers of glazing. Double glazing insulates almost twice as well as single glazing. Adding a third or fourth layer of glazing results in further improvement. In the 1970s, with rising concern over energy, triple-glazed and even quadruple-glazed windows entered the market. Some of these windows use glass only; others use thin plastic films as the inner glazing layer(s).

■ *Thickness of air space.* With double-glazed windows the air space between the panes of glass has a big effect on energy performance. A very thin air space does not insulate as well as a thicker air space because of the conductivity through that small space. During the 1970s a lot of window manufacturers increased the thickness of the air space in double-glazed windows from ¼" to ½" or more. If the air space is too wide, however, convection loops between the layers of glazing occur. Beyond about 1", you do not get any further gain in energy performance with thicker air spaces.

■ *Low-conductivity gas fill.* By substituting the air in a sealed insulated glass window for a denser, lower conductivity gas such as argon, heat loss can be reduced significantly. The largest window manufacturer in the country today, Andersen Windows, uses argon gas-fill in all of its insulated glass windows, and most major manufacturers offer argon gas-fill as an option. Other gases that have been or are being used in windows include carbon dioxide (CO_2), sulfur hexafluoride (SF_6), krypton (Kr), and argon-krypton mixtures.

■ *Tinted glass coatings.* Tinted glass and tinted window films have long been used in commercial buildings to reduce heat gain through windows. Improved, lightly tinted windows are becoming more common for the residential market in southern (cooling-dominated) climates. These new glazings reduce the solar heat gain without reducing visibility as much as older tinted glass and films.

LOW-E COATINGS

More than any other single improvement, the invention and commercial development of low-emissivity (low-e) coatings in the 1980s revolutionized window technology. Thin, transparent coatings of silver or tin oxide permit visible light to pass through, but they effectively reflect infrared heat radiation back into the room. This reduces heat loss through the windows in the winter.

A variety of low-e windows are now available for different climate zones and different applications in any particular location. Low-e windows with high solar gain coefficients are appropriate for northern climates where passive solar heating is advantageous, while "southern low-e" windows with low heat gain coefficients are appropriate in milder climates where summer cooling is more significant than winter heating.

The different properties of these low-e glazings sometimes make it advisable to choose different types of glazing for different sides of your house. We realize it may be difficult to find a builder or contractor interested in customizing window glazings for the four sides of your house. However, we encourage you to begin thinking about new windows from an informed energy perspective. For example, if you want to benefit from passive solar heating, for the south side choose windows containing a top-performing low-e glass with a high solar heat gain coefficient. On the north, install the lowest U-value windows you can afford. Or to keep things simpler, you can order the same glazings for the east-, west-, and north-facing windows.

Some window manufacturers now produce both "northern" and "southern" climate low-e products. But other window manufacturers may still offer just one type of low-e glazing as their standard, and charge extra for substitutions—if they provide options at all. So a decision to choose different glazings for windows of different orientations may require some extra shopping around. If you do order different glazings for your different windows, be sure to keep track of which windows have which type of glazing, because they will probably all look identical!

EDGE SPACERS

As window glazings have improved in performance, what happens at the edges of the windows becomes increasingly significant. The edge spacer is what holds the panes of glass apart and provides the airtight seal in an insulated glass window.

Traditionally, these have been hollow aluminum channels, usually filled with desiccant beads (which absorb small amounts of moisture that might get into the glazing unit).

Aluminum has extremely high conductivity. That didn't matter when the glazing did not insulate very well, but as better performing glazings were developed, proportionately more heat was lost through the edges.

Since about 1990, a number of improved edge spacers have come onto the market. Some are made of thin-walled steel and have a thermal

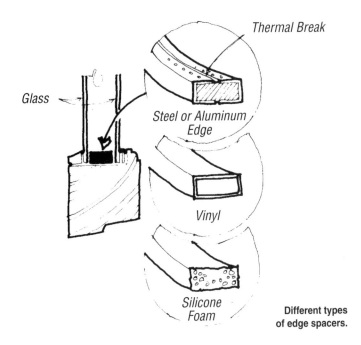

Different types
of edge spacers.

break. Others are made of silicone foam or butyl rubber. The net effect of improved edge spacers can be a 2-10% improvement in window energy performance, depending on the other performance characteristics of the window. With new edge spacers, however, pay particular attention to warranties against seal failure, which results in fogging and loss of any low-conductivity gas-fill. Choose windows with long warranties.

WINDOW FRAME AND SASH CONSTRUCTION

Window frame and sash construction has a big impact on energy performance. Wood is still the most common material in use, and it insulates reasonably well. Aluminum has been used extensively, particularly in the western part of the U.S., but unless a thermal break is incorporated into the design, aluminum frames conduct heat very rapidly and are therefore inefficient. Vinyl (PVC) windows are gaining in popularity, especially in the replacement market, and some vinyl frames insulated with fiberglass insulate better than wood.

Another important property is the airtightness of windows. Windows vary dramatically in how effectively they block infiltration. Airtightness is usually measured in cubic feet of air per linear foot of crack (cfm/ft) at specified testing conditions. The tightest windows have air leakage rates as low as 0.01 cfm/ft, and the industry standard is 0.37 cfm/ft. Most of the better windows have leakage rates in the range of 0.01 to 0.06 cfm/ft.

However, individual windows purchased off a shelf may have leakage rates as high as 1 cfm/ft! Consumers should recognize that the manufacturer's quality control at the factory and care during shipping can have a big impact on the window's air tightness at a site.

Furthermore, it is senseless to invest thousands of dollars in new windows only to have amateurs install them in your home. The high-rated performance of a window is meaningless if it is installed improperly with gaps and air leaks around the window. Be sure to have experienced contractors install your high-tech windows.

In general, casement and awning windows are tighter than double-hung and other sliding windows. This is because when a casement or awning window is closed it is pulled in against a compression-type gasket. Sliding windows have to use seals that permit the sash to slide, so they are rarely as airtight. You will find, though, that double-hung windows from a few manufacturers are tighter than casement windows

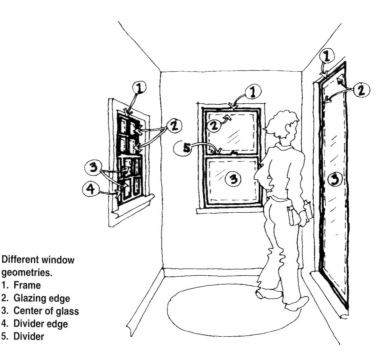

Different window
geometries.
1. Frame
2. Glazing edge
3. Center of glass
4. Divider edge
5. Divider

from others, so it makes a lot of sense to examine air leakage specifications carefully when selecting windows.

WINDOW DIMENSIONS

With high-performance windows, the dimensions of the windows have a big impact on total energy performance. This is because the glazing (glass, low-e coatings, gas fill, etc.) generally insulates a lot better than the edge of the window (edge spacer, sash, and frame). This aspect of window performance has led to a great deal of confusion. If you consider only the energy performance of the glazing itself (center-of-glass performance), you get a very high insulating value. But if you factor in the energy performance of the window edges to arrive at a unit or average performance value, the insulating value is significantly lower. The smaller the window in question, the more significant the effect of the window edges, because a small window has proportionately more edge area than glazing area. Windows with true divided lights have a great deal more edge area.

UNDERSTANDING NFRC WINDOW RATINGS

Beginning in 1993, some windows began carrying NFRC energy performance labels. The National Fenestration Rating Council is a nonprofit collaboration of window manufacturers, government agencies, and building trade associations, founded to establish a fair, accurate, and credible energy rating system for fenestration products (windows, doors, and skylights).

Before NFRC developed standardized rating procedures for heat loss through windows, there was little consistency in how manufacturers listed energy performance. For example, some manufacturers listed only the center-of-glass performance, while others averaged performance for the entire unit. Also, different methods were used by manufacturers to calculate energy performance, so you couldn't be sure that the performance numbers from one manufacturer could be fairly compared with those of another.

NFRC U-VALUE RATINGS

Windows that have been rated by NFRC-approved testing laboratories and certified by independent certification and inspection agencies carry labels with U-value (U-factor) ratings (see illustration). U-value is the measure of the amount of heat (in Btus) that moves through a square foot of window in an hour for every degree Fahrenheit difference in temperature across the window (Btu/ft_2Fhr). U-value is the inverse of R-value, which is familiar to many people as an indicator of levels of insulation. With wall and attic insulation, the higher the R-value the better the insulation. Since U-value is the *inverse* of R-value, **the lower the U-value rating, the better the overall insulating value of the window.**

The U-factor ratings listed on NFRC labels (and in the *NFRC Certified Products Directory*) are whole-window U-values. That is, they take into account heat loss through the glass, window edge, and window frame.

COMPARATIVE U-VALUE

Because the unit U-values differ depending on the size of the window and it is unrealistic to list a different rating for every window size offered, NFRC came up with two standard sizes for every window type (see Table 3.1). For example, the representative residential casement window is 24" × 48", and the representative nonresidential casement window is 30" × 60". Showing two different sizes illustrates how sensitive the overall window U-value is to window size. If the U-values are very different for the residential and nonresidential samples of a single product, that product's thermal performance is sensitive to the actual product size. When comparing different products, be sure to compare ratings for the same size.

TABLE 3.1

REPRESENTATIVE WINDOW SIZES FOR NFRC LABELING

	Size in inches (width × height)	
	Residential	Nonresidential
Vertical Slider	36 × 60	48 × 72
Horizontal Slider	60 × 36	72 × 48
Casement	24 × 48	30 × 60
Projecting (Awning)	48 × 24	40 × 40
Fixed (includes unusual shapes)	48 × 48	48 × 72
Swinging Door with Frame	38 × 82	40 × 96
Sliding Patio Doors with Frames	72 × 82	72 × 96
Skylights and Sloped Glazings	24 × 48	48 × 48
Greenhouse/Garden	60 × 36	72 × 48

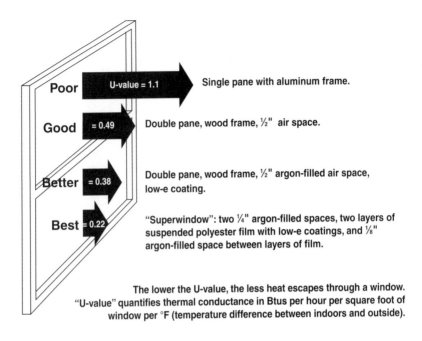

Poor — U-value = 1.1 — Single pane with aluminum frame.

Good — = 0.49 — Double pane, wood frame, ½" air space.

Better — = 0.38 — Double pane, wood frame, ½" argon-filled air space, low-e coating.

Best — = 0.22 — "Superwindow": two ¼" argon-filled spaces, two layers of suspended polyester film with low-e coatings, and ⅛" argon-filled space between layers of film.

The lower the U-value, the less heat escapes through a window. "U-value" quantifies thermal conductance in Btus per hour per square foot of window per °F (temperature difference between indoors and outside).

When a window manufacturer has its products certified, a great many different types and configurations of windows may be included. As defined by NFRC, a *product line* is a group of windows that have similar frame and operating characteristics (e.g., "Andersen Perma-Shield Casement windows"). Within each product line there may be various products. Differences among products include the number of layers and glazing, the use of or type of low-e coating, the type of gas-fill, edge spacer, etc.

UNDERSTANDING AN NFRC LABEL

NFRC labels describe the whole window U-factor (U-value), solar heat gain coefficient (SHGC), visible light transmittance, and air leakage for two representative sizes of windows.

NFRC SOLAR HEAT GAIN RATINGS

After heat loss, the most important energy property of glazings is how much solar heat gain they transmit. Solar heat gain can be beneficial—providing free passive solar heat during the winter months—or it can be a problem, resulting in overheating during the summer. The solar heat gain coefficient introduced earlier describes how much solar energy is transmitted through a window. An SHGC of 0.8 means that 80% of the solar energy hitting the window gets through. (The window

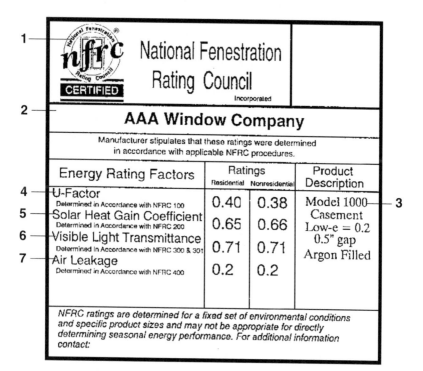

1. Official NFRC logo.
2. Window manufacturer.
3. Description of the individual product on which the label appears.
4. The certified U-factor (U-value) for the entire window; the *lower* the U-factor, the more efficient the window.
5. The certified solar heat gain coefficient (SHGC) for the entire window. A dimensionless number from 0 to 1.0; the higher the SHGC, the greater the solar energy that passes through the window system (glazing and frame).
6. The certified visible light transmittance (VT) value for the entire window. A dimensionless number from 0 to 1.0; the higher the VT, the greater the visible sunlight that passes through the window system (glazing and frame).
7. The certified air leakage rate. The air leakage rate is based on test conditions, and is given as the cubic feet (of air) per minute per foot of window edge (cfm/ft).

industry has used a similar term, "shading coefficient," but solar heat gain coefficient is becoming more common.)

NFRC has developed standard procedures for determining solar heat gain coefficients, and these ratings are listed on NFRC labels. These values make it easier to determine the suitability of windows for passive solar applications and the impact of windows on overheating and air conditioning loads.

But, as noted above, major window manufacturers have already started making low-e windows with different solar heat gain coefficients. Windows with high coefficients (above 0.70) are designed for colder climates, while windows with low coefficients are designed for hotter climates.

VISIBLE LIGHT TRANSMITTANCE

While SHGC describes the relative amount of solar energy that can pass through a window, the visible light transmittance is simply the relative amount of sunlight that can pass through. For example, tinted windows will have lower visible transmittance than clear windows, although the solar energy performance of tinted and clear windows might be similar. Also, a window with a large ratio of frame to glass will have a lower visible transmittance value than a window with a smaller ratio of frame to glass, since the frame will block visible sunlight.

NFRC INFILTRATION RATINGS

The third important energy property of windows is air leakage or air infiltration. Air leakage is already listed by many window manufacturers, in terms of cubic feet of air per minute per foot of crack. Air leakage is now included on NFRC labels and in the *NFRC Certified Products Directory*. The NFRC has adopted the same basic procedures for measuring air leakage that have been used by the industry in the past.

ENERGY STAR® WINDOWS

Windows, doors, and skylights qualifying for the ENERGY STAR® label must meet requirements tailored for the country's three broad climate regions: northern, central, and southern. Window and door products that qualify in the heating-dominated northern states must have a U-factor of 0.35 or below and skylights must have a U-factor of 0.45 or below. In the central climate, where both heat loss and solar heat gain are of some importance, qualifying products must have a U-factor of

0.40 or below and a SHGC of 0.55 or below. And in cooling-dominated southern states, qualifying products must have a U-factor of 0.75 or below and a SHGC of 0.40 or below.

In addition, each ENERGY STAR® window must carry the NFRC label, allowing comparisons of ENERGY STAR® qualified products on specific performance characteristics.

RECOMMENDATIONS

1. If you're shopping for new windows, look for the NFRC label as your guide to their energy performance. Remember that a window's ability to insulate is given by its U-value, and the amount of solar energy it transmits is given by its solar heat gain coefficient—the lower the U-value and SHGC, the more efficient the window. Compare ratings on ENERGY STAR®-labeled windows for your climate region.

2. Look for windows with these energy-saving features: double panes; low-e coatings; low-conductivity gas-fill between panes; and wood, vinyl, or fiberglass frames.

3. Select windows with low air leakage ratings—between 0.01 and 0.06 cfm/ft.

4. Consider different glazings for windows on different sides of your house to benefit from passive solar and maximize energy benefits. Install the lowest U-value windows you can afford on north-facing windows. Select windows with appropriate low-e coatings for your local climate on the east, west, and south sides of your house.

5. To maximize energy performance, choose windows with larger unbroken glazing areas instead of multi-pane or true-divided-light windows. Applied grills that simulate true-divided-light windows are fine; they do not reduce energy efficiency.

6. Choose windows with good warranties against the loss of the air seal. If the glazing seal is lost, not only will fogging occur, but also any low-conductivity gas between the layers of glass will immediately be lost.

7. To ensure that your new windows perform as well as they should, hire skilled contractors to install them.

CHAPTER 4
Heating Systems

Heating is the largest energy expense in most homes, accounting for around two-thirds of annual energy bills in colder parts of the country. Reducing your energy use for heating, therefore, provides your single most effective way to reduce your home's contribution to global environmental problems. Heating systems in the United States spew over a billion tons of CO_2 into the atmosphere each year and about 12% of the nation's sulfur dioxide and nitrogen oxides.

A combination of conservation efforts and a new high-efficiency heating system can often cut your pollution output and fuel bills in half. Just upgrading your furnace or boiler from 56% to 90% efficiency in an average cold-climate house will save 1.5 tons of CO_2 emissions if you heat with gas, or 2.5 tons of CO_2 emissions if you heat with oil.

Very basically, a heating system replaces heat that is lost through the shell of your house. How much energy your heating system requires to replace that lost heat depends on four factors: where the house is located (in colder places, the house will lose more heat); how big the house is; the energy efficiency of the house; and how energy-efficient the heating system is.

You can't do much about the first factor. As for size, the bigger the house, the more energy it will take to heat it, but you aren't likely to change the size of your house, except perhaps to make it bigger.

You can do something about energy efficiency levels. Improving your home's energy efficiency is an excellent idea and offers the greatest opportunity for savings in heating costs. Some of the most effective measures, such as rope-caulking windows, cost next to nothing, while others, such as insulating, can be both expensive and time-consuming. These measures are addressed in Chapter 2.

That leaves your heating system itself, the focus of this chapter. Whether or not you've buttoned up your house, you can probably save a great deal by upgrading your heating system, either by installing a new high-efficiency system or by boosting the efficiency of your present system. Both of these options are addressed in this chapter.

HEATING SYSTEM BASICS

In considering the various options for improving or replacing your heating system, it helps to know some of the lingo. A lot of confusing terms and concepts are going to get thrown around by salespeople or heating system technicians, and you don't want to get left behind. Central heating systems have three basic parts: the heating plant itself where fuel is converted into useful heat, a distribution system to deliver heat to where it is needed, and controls to regulate when the fans and pumps controlling it all turn on and off.

TYPES OF HEATING SYSTEMS

Gas and Oil Heating Systems

Gas- and oil-fired heating systems generate heat in a *furnace* or *boiler*. Many people are confused about the difference between furnaces and boilers. A furnace heats air that is blown through air ducts

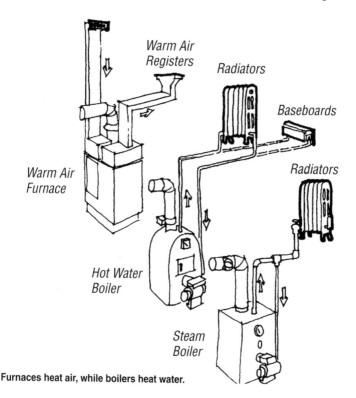

Furnaces heat air, while boilers heat water.

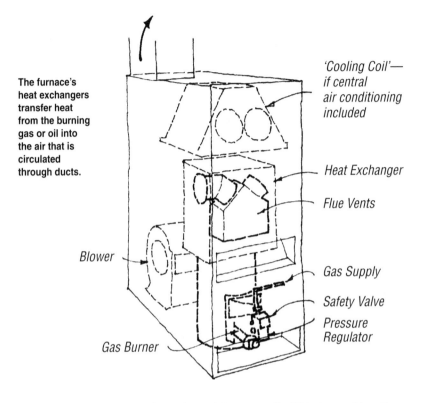

The furnace's heat exchangers transfer heat from the burning gas or oil into the air that is circulated through ducts.

'Cooling Coil'—if central air conditioning included

Heat Exchanger

Flue Vents

Blower

Gas Supply

Safety Valve

Pressure Regulator

Gas Burner

and delivered to rooms through *registers* or *grills*. This type of heating system is called a ducted warm-air or forced warm-air distribution system. A boiler heats water or steam that circulates through pipes to *radiators* or *baseboards*. Hot-water heating systems, sometimes called *hydronic systems*, are more common today than steam systems. Some hot-water systems circulate water through plastic tubing in the floor, a system called *radiant floor heating*. Furnaces and boilers are both widely available in gas- and oil-fired models.

Inside the furnace or boiler, fuel is sprayed into a combustion chamber where it is mixed with air and burned. The combustion products are vented out of the building through a flue pipe. The flames heat a metal box called a heat exchanger. In a furnace, air is heated in the heat exchanger, while in a boiler, water is heated in the heat exchanger. For hot-water systems, the water is heated to about 180°F; in steam systems, the water is boiled, creating steam.

Heating system controls regulate when the various components of the heating system turn on and off. The most important control from

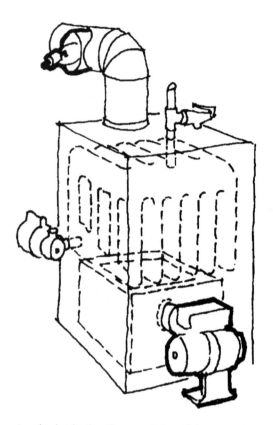

Boilers work
like furnaces,
except that
water instead
of air is heated.

your standpoint is the thermostat, which turns the system—or at least the distribution system—on and off to keep you comfortable. But there are other controls in a heating system, including aquastats, valves, vents, fan thermostats, and dampers. These control devices are described later in this chapter.

Efficiency. The efficiency of a gas or oil heating system is a measure of how effectively it converts fuel into useful heat. There are two types of efficiency. *Combustion efficiency* tells you the system's efficiency while it is running. Combustion efficiency is like the miles per gallon your car gets cruising along at 55 miles per hour on the highway.

A more accurate estimation of actual fuel use, though, is the *annual fuel utilization efficiency*, or AFUE. This is a measure of the system's efficiency that accounts for start-up and cool-down and other operating losses that occur in real operating conditions. AFUE is like your car mileage between fill-ups, including both highway driving and stop-and-go traffic.

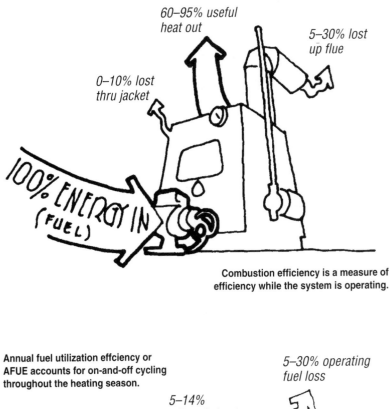

60–95% useful heat out

5–30% lost up flue

0–10% lost thru jacket

100% ENERGY IN (FUEL)

Combustion efficiency is a measure of efficiency while the system is operating.

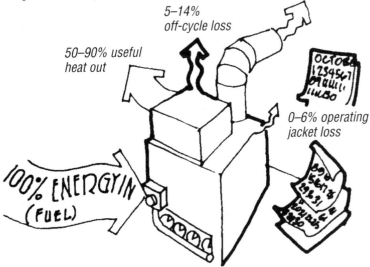

Annual fuel utilization effciency or AFUE accounts for on-and-off cycling throughout the heating season.

5–30% operating fuel loss

5–14% off-cycle loss

50–90% useful heat out

0–6% operating jacket loss

100% ENERGY IN (FUEL)

Electric Heat

There are two common types of electric heat: electric resistance heat and electric heat pumps. Electric resistance heat works by directly converting electric current into heat. Almost all of the energy in the electric current ends up as usable heat, but it is still pretty inefficient when the inefficiency of electricity generation by the power company and transmission losses are taken into account. Electric resistance heat is usually the most expensive form of heat and it is, therefore, seldom recommended. It will not be addressed further in this chapter.

Electric heat pumps operate on a quite different principle. Instead of directly producing heat from the electric current, they use

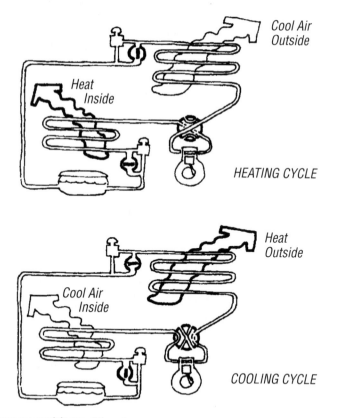

Heat pumps can work in two different modes: heating and cooling.

electricity to move heat from one place to another. They work in the same basic way as a refrigerator, using a special refrigerant fluid (HCFC-22 or newer HFC alternatives) that changes back and forth between liquid and vapor. In the heating mode, a heat pump extracts heat from outside the house and delivers it to the house. Like furnaces, most heat pumps work with forced warm-air delivery systems. *Air-source* heat pumps use the outside air as the heat source. *Ground-source* or "GeoExchange" heat pumps get their heat from underground, where temperatures are more constant year-round. Air-source heat pumps are far more common than ground-source heat pumps.

Because electricity in a heat pump is used to move heat rather than to generate it, the heat pump can deliver more energy than it consumes. The ratio of delivered energy to consumed energy is called the *coefficient of performance* or COP, with typical values ranging from 1.5 to 3.5. A better measure for comparing air-source heat pump performance is the *heating season performance factor* or HSPF. This is a ratio of the estimated seasonal heating output (Btus) divided by the seasonal power consumption (watts).

One of the advantages of heat pumps is that they can be used for cooling during the summer months. In the cooling mode, the cycle is reversed: heat is taken from the house and dumped to the outside air or the ground. There will be more discussion of heat pumps in the section on selecting new heating systems.

Wood Heat

Wood heating can make a great deal of sense in rural areas if you don't mind stacking wood and stoking the stove or furnace. Wood prices are generally lower than gas, oil, or electricity. If you cut your own wood, the savings can be large. Pollutants from wood burning have been a problem in some parts of the country, causing the U.S. Environmental Protection Agency (EPA) to implement regulations that govern pollution emissions from wood stoves. As a result, new models are quite clean-burning. This edition of the *Consumer Guide to Home Energy Savings* does not include detailed information on wood stoves.

If possible, provide a combustion air source for wood stoves.

UNVENTED GAS HEATERS: A BAD IDEA

Unvented gas space heaters and unvented gas fireplaces are growing in popularity, but we strongly discourage their use for health and safety reasons. Known as "vent-free" gas heating appliances by manufacturers, they include wall-mounted and free-standing heaters as well as open-flame gas fireplaces with ceramic logs.

Manufacturers claim that because the products' combustion efficiency is very high, they are safe for building occupants. However, these efficiency claims are valid only under certain conditions, including keeping a nearby window open "an inch or two" for adequate fresh air—which defeats the purpose of supplemental heat. Kerosene heaters, which were popular in the 1970s, are similar except the fuel is liquid kerosene instead of natural gas. Most states have banned unvented kerosene heaters from residential use because of the same dangers discussed here for vent-free gas appliances. Combustion by-products, unavoidable indoors from unvented heaters, include nitrogen oxides (eye, nose, and throat irritants), carbon monoxide (an asphyxiant), and a lot of water vapor, which can cause condensation, mildew, or rot in wall and ceiling cavities.

Another problem is oxygen depletion. In the absence of a flame, air is 21% oxygen. Since flames consume oxygen much more strongly than human lungs, unvented gas heaters must be equipped with oxygen depletion sensors, which automatically shut off the gas to the heaters if oxygen falls below 18% of room air. Because of the hazards posed by vent-free appliances, five states (California, Minnesota, Massachusetts, Montana, and Alaska) prohibit their use in homes, and many cities in the United States and Canada have banned them as well.

Rather than purchasing an unvented gas heater, we recommend consumers stay warm and healthy by reducing heat loss from their home as described in Chapter 2, and improving their central heating system as described in this chapter.

SHOULD I REPLACE MY EXISTING HEATING SYSTEM?

This can be a difficult question. If you heat with electric resistance heat, rising electricity prices may force you to switch to a gas, oil, or heat pump system that is more affordable. If you currently have a gas- or oil-fired furnace or boiler, the decision to replace it depends on its age, condition, and performance.

If your furnace or boiler is old, worn out, inefficient, or significantly oversized, the simplest solution is to replace it with a modern high-efficiency model. Old coal burners that were switched over to oil or gas are prime candidates for replacement, as are gas furnaces without electronic (pilotless) ignition or a way to limit the flow of heated air up the chimney when the heating system is off (vent dampers or induced draft fan).

A typical heating system will last about 25 years, though some boilers can last twice that long. Your heating system technician or energy auditor may be able to help you evaluate your existing system and decide whether replacement is a good idea. If you're not going to replace it, see the sections on maintenance and modifications for ways to boost its efficiency and performance.

If you know how efficient your existing heating system is (AFUE), it's pretty easy to calculate the savings you will get by replacing it. The chart below will help you determine potential savings resulting from replacement of your existing system. Your heating service technician or energy auditor may be able to help determine the AFUE of your present system. If you were only provided with the combustion efficiency, you can estimate the AFUE by multiplying the combustion efficiency by 0.85. The numbers in the chart assume that both the old and new systems are sized properly; savings will be greater than indicated if the old system is too large.

DOLLAR SAVINGS PER $100 OF ANNUAL FUEL COST

		AFUE of new system								
		55%	60%	65%	70%	75%	80%	85%	90%	95%
AFUE	50%	$9	$16	$23	$38	$33	$37	$41	$44	$47
of	55%		8	5	21	26	31	35	38	42
existing	60%			7	14	20	25	29	33	37
system	65%				7	13	18	23	27	32
	70%					6	12	17	22	26
	75%						6	11	16	21
	80%							5	11	16
	85%								5	11

To determine savings, find the horizontal row corresponding to the old system's AFUE, then choose the number from that row that is in the vertical column corresponding to the new system's AFUE.

That number is the projected dollar savings per hundred dollars of existing fuel bills. For example, if your present AFUE is 65% and you plan to install a high-efficiency natural gas system with an AFUE of 90%, then the projected saving is $27 per $100. If, say, your annual fuel bill is $1,300, then the total yearly savings should be about $27 × 13 = $351.

That's a lot of money to save each year, especially when you consider the expected lifetime of a heating system, but it still doesn't answer the question of whether replacing the system is a good investment. To answer that, you can calculate the first year return on investment (ROI). The equation is as follows:

$$ROI = \text{first year savings} \div \text{installed cost}$$
$$ROI = \$351 \div \$2,500 = 0.14 = 14\%$$

A 14% return is pretty good—much higher than what you receive from a savings account or certificate of deposit. Plus, unlike most other investments, energy conservation investments are tax free. If fuel prices go up, the annual savings and return on investment also go up. For example, if fuel prices increase 30%, the annual savings in this example increases to $456, and the return on investment increases to 18%.

SELECTING A NEW HEATING SYSTEM

Fuel Options

If you're going to replace your existing heating system, you have a number of decisions to make. The first is what type of fuel to use. It usually makes sense to stick with the fuel your old system used, because the storage tank and necessary fuel supply piping will already be in place. But sometimes it makes sense to switch fuels. If you currently have electric resistance heat, for example, switching to gas or oil will probably save you money. Even electric heat pumps are more expensive than oil or gas heat in most of the country, though in warmer areas the use of a heat pump for cooling can help justify its use.

Switching from oil to gas or vice-versa based on current prices in your area, though, can be risky. Even though gas is usually cheaper today doesn't mean it will stay cheaper now that the gas industry is deregulated. Similarly, fuel oil prices fluctuate widely, as we have seen in recent years.

Switching from One Distribution System to Another

In some cases when replacing a heating system, you might want to change from one type of distribution system to another. If you have steam heat, for example, switching to hot water may make sense. Some steam piping systems can be retrofitted for hot-water heat. Discuss this possibility with the heating technicians you talk with. If you have a forced warm-air or hot-water system, though, it rarely makes sense to switch to something else. The cost of adding either new ducts or new piping will probably make the project prohibitively expensive.

Sizing the System

Before you actually start shopping for a new furnace or boiler, it pays to figure out how large a system you need. A system that is too large wastes fuel and money because it keeps cycling on and off. It only runs at peak efficiency for short periods of time and spends most of the time either warming up or cooling down. Many existing systems that were installed in the 1950s and 1960s are much too large. It is not uncommon for a heating system to be two or three times larger than necessary.

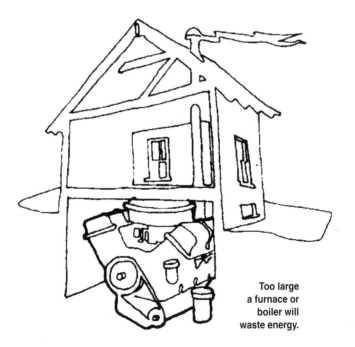

Too large
a furnace or
boiler will
waste energy.

efficient, your heating system should have to run almost on the coldest day of the year to keep your house 70°F. out how big a system you need by checking the size of the old system. If your heating contractor wants to do this, take your business elsewhere. A heat loss analysis is the only way to determine the proper size of a new heating system. (Steam systems are an exception: the boiler should be sized to the radiators.) A heat loss analysis can be done by your heating contractor or energy auditor; it should include measurements of wall, ceiling, floor, and window areas and account for insulation levels and weatherization features, including any energy improvements you have made. From the heat loss analysis, you need to know the peak hourly heating demand in Btu/hr on the coldest expected day of the year in your area (British thermal units or Btus are a measure of heat; one Btu is approximately the amount of heat released by burning a wooden kitchen match).

All heating systems are rated according to their Btu/hr output. A new heating system should be sized no more than 25% over the peak hourly heating demand. For example, if your home's peak hourly heating demand is calculated to be 60,000 Btu/hr, you should select a heating system with a heating output between 60,000 and 72,000 Btu/hr. Ask to see a copy of the heat loss and system sizing calculations so that you can make sure they were thoroughly done.

Several manufacturers have combined water heating and space heating functions in one integrated piece of equipment. If you are considering replacing both your furnace and your water heater, such a combination system may be advantageous to you. These are discussed in more detail in Chapter 6, under **Indirect Water Heaters** and **Advanced Heating Systems with Integrated Water Heaters**.

Shopping for Gas- and Oil-Fired Heating Systems

Because a new system will cost $1,000 or more, you should shop carefully. You will need to decide on a fuel, and you will need to balance cost, efficiency, contractor reliability, warranty, and other considerations.

If you live in a cold climate, it usually makes sense to invest the extra few hundred dollars for the highest efficiency system available. In milder climates with lower annual heating costs, the extra investment required to go from 80% to 90-95% efficiency may be hard to justify.

When shopping for high-efficiency furnaces and boilers, look for dependability. Buy a system with a good warranty and a reputable company to back it up. Most quality furnaces have a limited lifetime warranty on the heat exchanger, and most hot-water boilers come with a 20-year warranty. The controls should have at least a one-year warranty. If the manufacturer is a relatively new company, ask the dealer about the company's reputation. Ask for the names of several homeowners in your area with their systems in place, and call to see if the homeowners are satisfied with performance and service.

With gas and oil systems, specify sealed combustion. Sealed combustion appliances bring outside air directly into the burner and exhaust flue gases directly to the outside, without the need for a draft hood or damper. They generally burn more efficiently, and they pose no risk of introducing dangerous combustion gases into your house. With non-sealed-combustion appliances, back-drafting of combustion gases can be a big problem.

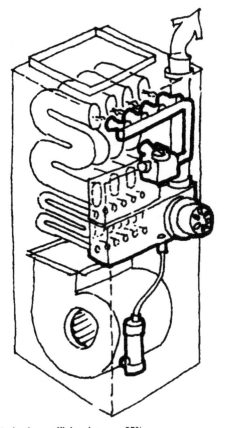

The best furnaces today have efficiencies over 95%.

High-efficiency systems—over about 85%—generally achieve that high efficiency in part by reclaiming most of the heat in the exhaust gases, including water vapor, that would otherwise escape up the chimney. The temperature of exhaust gases from a condensing gas furnace may be as low as 100°F—compared with 400°F or so for a conventional model. By sending less heat up the chimney, you deliver more heat to your house and keep more money in your pocket.

When exhaust gases cool below 212°F, water vapor in the exhaust condenses into water. The good news is that this releases additional heat, further boosting efficiency. The bad news is that the water is somewhat acidic and can corrode a conventional chimney. If the system will vent into an existing chimney that isn't lined, consider having a lining installed. Or better still, don't use a chimney at all; most of these very high-efficiency systems vent the exhaust gases to the outside through a plastic or stainless steel pipe that is installed through the basement wall. To handle the water produced, most condensing systems have condensate drains as well as exhaust drains.

Choosing a Heating System Contractor

You'll be spending a lot of money to have a new furnace or boiler installed—anywhere from about $1,500 for a simple furnace installation to over $4,500 for a very complicated boiler installation. Any modifications to the distribution system will add significantly to the cost. To get the best deal, you should get bids for the price of equipment and installation from several contractors. It isn't unusual for bids to differ as much as $1,000, so get at least two or three bids.

When evaluating bids, look at prices, but also look at what you get for the price: quality, energy savings, and warranty. If you are not familiar with the contractor, ask for customer references and follow up on them. Make sure the contractor is fully bonded and insured. If you are putting in a very high-efficiency furnace or boiler, ask the contractor if he or she has had any special training in this type of installation. If you think your old heating system is covered with asbestos insulation, make sure the contractor knows about it and will follow the proper procedures to deal with the asbestos. See Appendix 5 for more information on selecting a contractor.

Installation and Quality Control

A new heating system must be installed properly to ensure safe and efficient operation. With oil-fired furnaces and boilers, the system should be tuned and a combustion efficiency test performed after in-

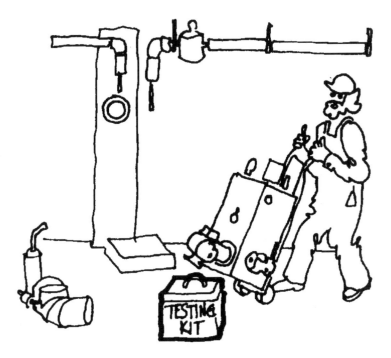

Make sure your heating system installer has
experience with high-efficiency systems.

stallation. Make sure the combustion efficiency meets the specifications (a calculation to convert from combustion efficiency to AFUE may
need to be made). Gas furnaces and boilers come pre-adjusted from
the factory and don't require a combustion efficiency test.

With both oil and gas systems, look for a neat installation. Mistakes
are more likely if an installation is sloppy. If the furnace or boiler is very
efficient (AFUE over about 85%), make sure that a plastic or stainless
steel exhaust pipe for flue gases and a condensate drain have both
been installed. The drain tube should run directly into a floor drain or
into a bucket, which you will need to empty periodically.

THE MOST EFFICIENT FURNACES AND BOILERS

The most efficient furnaces and boilers on the market are listed in
the following tables. The efficiency ranges vary according to the type of
system. Gas furnaces generally have the highest efficiencies of any
systems available. All the models listed below have AFUE ratings over

93%, and the best exceed 96%. The best gas boilers achieve efficiencies in the upper-80% range, and one manufacturer, Dunkirk, has introduced a 95% efficient boiler called the "Quantum Leap." The best oil furnaces and boilers have efficiencies in the mid- to upper-80% range. In general, steam boilers are somewhat less efficient than hot-water boilers, although the best systems still have efficiencies over 80%. Products are listed in several different heating capacity groupings for convenience in finding what's best for your situation.

To qualify for an ENERGY STAR® label, a furnace (oil or gas) must have an AFUE of 90% or higher, and a boiler (oil or gas) much have an AFUE of at least 85%. For more information about ENERGY STAR® heating equipment, including special ENERGY STAR® financing options, please call the toll-free ENERGY STAR® hotline: 1-888-782-7937.

The amount of electricity drawn by a furnace to power its motors and blow air through the house is often overlooked, since we focus on the natural gas or fuel oil burned to make heat. However, the electric energy demand by the furnace can be considerable—over 1,200 kWh/year for some models, adding up to $100 to your yearly electric bill. How much electricity a furnace uses depends on the design and configuration of the unit, and the efficiency of the motors and fans.

In addition to the AFUE ratings for the most efficient furnaces, we list below the estimated amount of electricity each will use in an average year. This power consumption is not factored into the AFUE ratings, so it is another item for consideration when choosing a new furnace. The actual amount of electricity used by any furnace will vary with your local weather and house characteristics, such as the efficiency of your ducts.

MOST EFFICIENT GAS FURNACES				
Manufacturer	Model	Btu/hr	Electric Energy (kWh/yr)	AFUE (%)
26,000 - 42,000 Btu/hr heating capacity				
Bryant	355MAV042040	38,000	105	96.6
Carrier	58MVP040-14	38,000	105	96.6
Lennox	G21Q3-40-	39,000	261	96.2
Bryant	355MAV042040**	37,000	106	96.1
Carrier	58MVP040-14**	37,000	106	96.1
Bryant	350MAV036040	38,000	458	95.5
Carrier	58MXA040-12	38,000	458	95.5
Bryant	355MAV042040*	37,000	107	95.0
Carrier	58MVP040-14*	37,000	107	95.0
Bryant	350MAV036040**	38,000	460	94.9
Carrier	58MXA040-12**	38,000	460	94.9
Bryant	350MAV024040	38,000	317	94.3

MOST EFFICIENT GAS FURNACES (cont.)

Manufacturer	Model	Btu/hr	Electric Energy (kWh/yr)	AFUE (%)
26,000 - 42,000 Btu/hr heating capacity (cont.)				
Carrier	58MXA040-08	38,000	317	94.3
Bryant	355MAV042060-L	37,000	68	94.1
Carrier	58MVP060-14-L	37,000	68	94.1
43,000 - 59,000 Btu/hr heating capacity				
Amana	GUX045X30B	44,000	506	96.9
Amana	GUD045X30B	43,000	538	95.6
Lennox	GSR21Q4-50-	48,000	390	95.3
Amana	GUC045X30B	43,000	500	95.3
Armstrong	GU95A045V12	43,000	223	95.0
Lennox	GSR21Q3-50-	48,000	418	94.8
Thermo-Products	CHB-50	47,000	294	94.5
Lennox	G21V3-60-	57,000	105	94.3
Bryant	355MAV060080-L	49,000	84	94.1
Carrier	58MVP080-20-L	49,000	84	94.1
Bryant	355MAV042060	57,000	95	94.1
Carrier	58MVP060-14	57,000	95	94.1
Lennox	G21Q3-60-	56,000	317	94.1
Lennox	G21Q4-60-	57,000	415	94.1
Ducane	FPAA050D13CN	47,000	856	94.0
Evcon Industries	CGH05012D	47,000	856	94.0
Int'l Comfort/Sears	NTVM050FF***	47,000	149	94.0
Int'l Comfort/Sears	VNK050F12***	47,000	149	94.0
Intertherm/Miller	G2RS050C12	47,000	856	94.0
Olsen Technology	GTH 050	47,000	484	94.0
Rheem/Ruud	*GFD-06*MCKS	58,000	313	94.0
Unitary Products Group	P-HDD12N04701B	47,000	856	94.0
Unitary Products Group	G9D0612UPB11	57,000	631	94.0
Weatherking	WGFD-06*MCKS	58,000	313	94.0
York	P1XUB12N0551	57,000	631	94.0
60,000 - 76,000 Btu/hr heating capacity				
Amana	GUX070X30B	67,000	614	95.9
Amana	GUX070X40B	67,000	634	95.9
Armstrong	GU95A067V14	64,000	200	95.0
Lennox	GSR21V5-80-	76,000	242	94.6
Lennox	GSR21V3-80-	76,000	172	94.5
Lennox	G21V3-80-	76,000	190	94.5
Amana	GDC070X30B	66,000	624	94.4
Evcon Industries	G9T08016UPC13	76,000	795	94.3
Unitary Products Group	G9T08016UPC13	76,000	795	94.3
York	P3URC16N07501	76,000	795	94.3
Amana	GDC070X40B	66,000	747	94.2

MOST EFFICIENT GAS FURNACES (cont.)

Manufacturer	Model	Btu/hr	Electric Energy (kWh/yr)	AFUE (%)
60,000 - 76,000 Btu/hr heating capacity (cont.)				
Bryant	355MAV060080	75,000	103	94.1
Carrier	58MVP080-20	75,000	103	94.1
Bryant	355MAV042080	75,000	118	94.1
Carrier	58MVP080-14	75,000	118	94.1
Bryant	355MAV060100-L	61,000	161	94.1
Bryant	355MAV060120-L	73,000	161	94.1
Carrier	58MVP100-20-L	61,000	161	94.1
Carrier	58MVP120-20-L	73,000	161	94.1
Lennox	GSR21Q4/5-80-	76,000	600	94.1
77,000 - 93,000 Btu/hr heating capacity				
Amana	GUX090X50B	86,000	735	95.8
Amana	GUX090X35B	86,000	735	95.8
Armstrong	GU95A090V16	86,000	284	95.0
Armstrong	GU95A090V20	86,000	285	95.0
Olsen Technology	GTH 085	81,000	815	95.0
Amana	GUD090X50B	85,000	652	94.6
Amana	GUC090X35B	85,000	695	94.4
Amana	GUC090X50B	85,000	695	94.4
Amana	GDC090X50B	87,000	771	94.4
Amana	GDC090X40B	87,000	816	94.2
Amana	GCD090X40B	86,000	840	94.0
Rheem/Ruud	*GFD-09*ZCMS	86,000	359	94.0
Weatherking	WGFD-09*ZCMS	86,000	359	94.0
Bryant	355MAV060100**	93,000	167	93.7
Carrier	58MVP100-20**	93,000	167	93.7
Amana	GUD090X35B	84,000	645	93.7
Dayton	MGRA-09EZAJS	84,000	848	93.5
Heat Controller	GLUA90-E5(N,NX)	84,000	848	93.5
Unitary Products Group	G9D10014UPC11	93,000	809	93.5
York	P1XUC14N09501	93,000	809	93.5
Lennox	G32Q4/5-100-*	93,000	987	93.2
94,000 - 110,000 Btu/hr heating capacity				
Amana	GUX115X50B	110,000	827	95.3
Armstrong	GU95A112V20	107,000	321	95.0
Olsen Technology	GTH 100	95,000	913	95.0
Lennox	G21V5-100-	95,000	460	94.5
Lennox	G21Q4/5-100-	95,000	794	94.5
Bryant	355MAV060100	94,000	225	94.1
Carrier	58MVP100-20	94,000	225	94.1
Int'l Comfort/Sears	NTVM100HJ***	94,000	269	94.0
Int'l Comfort/Sears	VNK100J16***	94,000	269	94.0
Lennox	G21Q3-100-	94,000	508	94.0

MOST EFFICIENT GAS FURNACES (cont.)

Manufacturer	Model	Btu/hr	Electric Energy (kWh/yr)	AFUE (%)
94,000 - 110,000 Btu/hr heating capacity (cont.)				
Rheem/Ruud	*GFD-10*ZCMS	100,000	537	94.0
Thermo-Products	CHB-100	94,000	521	94.0
Unitary Products Group	G9D10020UPC11	95,000	1358	94.0
Weatherking	WGFD-10*ZCMS	100,000	537	94.0
York	P1XUC20N09501	95,000	1358	94.0
Amana	GUC115X50B	108,000	800	93.7
Lennox	G32Q3/4-100-*	94,000	721	93.7
Amana	GUD115X50B	108,000	777	93.6
Nordyne	G6RD100*-14	95,000	1160	93.5
Rheem/Ruud	*GRA-09(E,N)ZAJS	84,000	848	93.5
Weatherking	WGRA-09(E,N)ZAJS	84,000	848	93.5
111,000 - 130,000 Btu/hr heating capacity				
Armstrong	GU95A125V20	119,000	441	94.5
Lennox	G32Q4/5-125-*	118,000	955	94.2
Lennox	G32V5-125-*	118,000	547	94.2
Bryant	355MAV060120	113,000	258	94.1
Carrier	58MVP120-20	113,000	258	94.1
Int'l Comfort/Sears	NTVM125KN***	118,000	252	94.0
Int'l Comfort/Sears	VNK125N20***	118,000	252	94.0
Rheem/Ruud	*GFD-12*RCMS	115,000	466	94.0
Unitary Products Group	G9D12020UPD11	112,000	1493	94.0
York	P1XUD20N11201	112,000	1493	94.0
Weatherking	WGFD-12*RCMS	115,000	466	94.0

MOST EFFICIENT GAS BOILERS – HOT WATER

Manufacturer	Model	Btu/hr	AFUE (%)
43,000 - 59,000 Btu/hr heating capacity			
Dunkirk	QL-50	48,000	95.0
Ultimate	Q90-50	48,000	90.0
Glowcore	GB060A	53,000	89.0
Burnham	RV3(N/P)S(L/P)-	55,000	88.0
Teledyne Laars	H(W)(P)-M2-60	53,000	87.0
Trianco-Heatmaker	H(W)(P)-M2-60	53,000	87.0
Unus	BN60	53,000	87.0
Utica	SC-3	44,000	87.0
Weil-McLain	AHE-60	51,000	85.5
Peerless	PDE-03-HS	56,000	85.0
60,000 - 76,000 Btu/hr heating capacity			
Dunkirk	QL-75	71,000	95.0
Dunkirk	Q90-75	71,000	90.0
Weil-McLain	GV-3	61,000	87.5
Slant/Fin	KC90	75,000	85.8

MOST EFFICIENT GAS BOILERS – HOT WATER (cont.)

Manufacturer	Model	Btu/hr	AFUE (%)
60,000 - 76,000 Btu/hr heating capacity (cont.)			
Dunkirk	DPFG-3T-70	60,000	85.0
Ultimate	PFG-3T-70	60,000	85.0
77,000 - 93,000 Btu/hr heating capacity			
Hydrotherm	AM-100	90,000	90.4
Glowcore	GB090A	80,000	88.8
Burnham	RV4(N/P)S(L/P)-	84,000	87.6
Weil-McLain	GV-4	92,000	87.3
Utica	SC-4	87,000	87.0
Teledyne Laars	H(W)(P)-M2-100	86,000	85.8
Trianco-Heatmaker	H(W)(P)-M2-100	86,000	85.8
Unus	BN100	86,000	85.8
Slant/Fin	CB-90	77,000	85.4
Axeman-Anderson	GL-91D	78,000	85.3
Dunkirk	DPFG-4T-91	78,000	85.0
Peerless	PDE-04-HS	84,000	85.0
Ultimate	PFG-4T-91	78,000	85.0
94,000 - 110,000 Btu/hr heating capacity			
Dunkirk	QL-100	95,000	95.0
Dunkirk	Q90-100	95,000	90.0
Teledyne Laars	E*P 110	94,100	85.5
Dunkirk	DPFG-5T-112	96,000	85.0
Ultimate	PFG-5T-112	96,000	85.0
111,000 - 140,000 Btu/hr heating capacity			
Hydrotherm	AM-150	135,000	90.6
Glowcore	GB130A	115,000	88.8
Teledyne Laars	CB-M2-150	135,000	88.6
Trianco-Heatmaker	CB-M2-150	135,000	88.6
Burnham	RV5(N/P)S(L/P)-	114,000	87.4
Weil-McLain	GV-5	122,000	87.2
Utica	SC-5	122,000	87.0
Slant/Fin	CB-135	116,000	85.2
Peerless	PDE-05-HS	111,000	85.0
Dunkirk	DPFG-6T-133	114,000	85.0
Ultimate	PFG-6T-133	114,000	85.0
Viessmann	RN-140	119,000	85.0

MOST EFFICIENT GAS BOILERS – STEAM

Manufacturer	Model	Btu/hr	AFUE (%)
77,000 - 93,000 Btu/hr heating capacity			
Burnham	IN4PVNI-	87,000	82.2
Weil-McLain	EG-35-PIDN-S	81,000	82.0
Burnham	IN4(S/L)(N/P)I-	87,000	82.0
New Yorker	CGS40A(N/P)I-	86,000	81.4
Hydrotherm	VS-110BS-PV	89,000	81.3
Smith Cast Iron	GB200-S-4L-INTD	89,000	81.3
Slant/Fin	GXH-105EDPZ	86,000	81.0
Utica	PEG-112BID	90,000	81.0
94,000 - 110,000 Btu/hr heating capacity			
Hydrotherm	VS-135BS-PV	110,000	81.5
Smith Cast Iron	GB200-S-4H-INTD	110,000	81.5
Weil-McLain	EG-40-PIDN-S	101,000	81.3
Slant/Fin	GXH-125EDPZ	104,000	81.0
111,000 - 127,000 Btu/hr heating capacity			
Burnham	IN5PVNI-	116,000	82.2
Burnham	IN5(S/L)(N/P)I-	115,000	82.0
New Yorker	CGS50A(N/P)I-	115,000	81.4
Weil-McLain	EG-45-PIDN-S	122,000	81.4
Slant/Fin	GXH-150EDPZ	120,000	81.1
Utica	PEG-150BID	120,000	81.0
128,000 - 144,000 Btu/hr heating capacity			
Burnham	IN6(S/L)(N/P)I-	144,000	82.1
New Yorker	CGS60A(N/P)I-	143,000	81.4
Hydrotherm	VGA-175BS-PV	142,000	81.3
Smith Cast Iron	GB250-S-5L-INTD	142,000	81.3
Hydrotherm	VS-165BS-PV	134,000	81.2
Smith Cast Iron	GB200-S-5-INTD	134,000	81.2
Slant/Fin	GXH-170EDPZ	137,000	81.1
Weil-McLain	EG-50-PIDN-S	142,000	81.1
145,000 - 178,000 Btu/hr heating capacity			
Burnham	IN6PVNI-	145,000	82.2
Burnham	IN7(S/L)(N/P)I-	173,000	82.1
Hydrotherm	VGA-200BS-PV	163,000	81.6
Smith	GB250-S-5H-INTD	163,000	81.6
New Yorker	CGS70A(N/P)I-	172,000	81.5
Slant/Fin	GXH-210EDPZ	172,000	81.2
Crown	JBF-72-EID	171,000	81.1
Dunkirk	PSB-7D	171,000	81.1
Slant/Fin	GXH-190EDPZ	155,000	81.1
Utica	PEG-187BID	151,000	81.0

MOST EFFICIENT OIL FURNACES

Manufacturer	Model	Btu/hr	Electric Energy (kWh/yr)	AFUE (%)
43,000 - 59,000 Btu/hr heating capacity				
Thermo Products	OL2-56-V	57,000	754	85.6
Olsen Technology	BML-60	59,000	665	85.0
DMO Industries	BML-60	59,000	1124	85.0
Armstrong	LUF80B57D10R	57,000	960	84.0
Armstrong	LBF80B57D10R	57,000	850	84.0
Armstrong	LBR80B57D10R	57,000	785	84.0
Oneida Royal	H056A-3,3B	59,000	500	84.0
Oneida Royal	LOR56A-3,3B	59,000	539	84.0
Oneida Royal	L056A-3,3B	59,000	531	84.0
Thermo Products	OC2-56-V	57,000	471	84.0
Thermo Products	OH2-56-V	57,000	455	83.7
Int'l Comfort Products	LUOV090*	59,000	566	83.2
Sears	LUOV090*	59,000	566	83.2
60,000 - 76,000 Btu/hr heating capacity				
Metzger	CFO-85	71,000	1029	85.9
Metzger	*BO-10A-85*	71,000	1029	85.9
Williamson	WCFO-85	71,000	1029	85.9
Williamson	WHBO-10A-85*	71,000	1029	85.9
Williamson	WLBO-10A-85*	71,000	1029	85.9
Bard	FH085D36* *	74,000	634	85.2
DMO Industries	BML-80	76,000	1187	85.1
Olsen Technology	BML-80	76,000	725	85.1
Newmac	NL2 77	77,000	688	84.8
Newmac	LFR 72-81	73,000	693	84.6
Newmac	LFR 61-66	62,000	547	84.5
Olsen Technology	HML 60B	61,000	1216	84.4
Bard	FLF085D36* *	74,000	652	84.2
Oneida Royal	LOR73A-3,3B	76,000	496	84.0
77,000 - 93,000 Btu/hr heating capacity				
Bard	FC085D36**	77,000	773	86.6
Bard	FC085D36***	88,000	746	86.1
Thermo Products	OC5-85-V	83,000	561	86.0
Thermo Products	OH5-85M	91,000	674	86.0
Thermo Products	OL5-85M	85,000	587	86.0
Metzger	CFO-105	88,000	1046	85.4
Metzger	*BO-10A-105*	88,000	1046	85.4
Williamson	WCFO-105	88,000	1046	85.4
Williamson	WHBO-10A-105*	88,000	1046	85.4
Williamson	WLBO-10A-105*	88,000	1046	85.4
Bard	FLR085D36***	86,000	796	85.0
Thermo Products	OH5-85-V	83,000	608	85.0
Bard	FLF085D36***	85,000	799	84.7
Columbia	CLB85*	88,000	721	84.6
Columbia	CLBD85*	88,000	721	84.6
Boyertown	LB85*	88,000	721	84.6

MOST EFFICIENT OIL FURNACES (cont.)

Manufacturer	Model	Btu/hr	Electric Energy (kWh/yr)	AFUE (%)
77,000 - 93,000 Btu/hr heating capacity (cont.)				
Boyertown	LBD85*	88,000	721	84.6
Newmac	NL2 89	89,000	638	84.6
Victa Hytemp/Healthaire	800CF/R	82,000	450	84.3
Bard	FH085D36***	84,000	698	84.2
94,000 - 119,000 Btu/hr heating capacity				
Bard	FLF110D48* *	97,000	1023	85.7
Metzger	CFO-120	100,000	1108	85.0
Metzger	*BO-10A-120*	100,000	1108	85.0
Metzger	*BO-12A-140*	116,000	1113	85.0
Thermo Products	OH11-105M	107,000	610	85.0
Williamson	WCFO-120	100,000	1108	85.0
Williamson	W*BO-10A-120*	100,000	1108	85.0
Williamson	W*BO-12A-140*	116,000	1113	85.0
Bard	FLR110D48* *	98,000	1034	84.8
Newmac	NL2 101	101,000	743	84.6
Thermo Products	OL11-105-V	101,000	897	84.6
Bard	FLF110D48***	113,000	1157	84.5
Bard	FH110D60* *	96,000	854	84.2
Bard	FH110D48* *	96,000	823	84.2
Bard	FLR110D48***	114,000	1161	84.2
Metzger	CFO-140	116,000	1113	84.2
Williamson	WCFO-140	116,000	1113	84.2
Columbia	CLB115*	115,000	1035	83.9
Columbia	CLBD115*	115,000	1035	83.9
Boyertown	LB115*	115,000	1035	83.9
Boyertown	LBD115*	115,000	1035	83.9

MOST EFFICIENT OIL BOILERS - HOT WATER

Manufacturer	Model	Btu/hr	AFUE (%)
60,000 - 76,000 Btu/hr heating capacity			
Dunkirk	DPFO-3T-.50	62,000	87.6
Melvin	CSTO-3T-.50	62,000	87.6
Ultimate	PFO-3T-.50	62,000	87.6
Dunkirk	DPFO-4-.60	74,000	87.5
Dunkirk	PFO-4-.60	74,000	87.5
Melvin	CSTO-4-.60	74,000	87.5
Ultimate	PFO-4-.60	74,000	87.5
Viessmann	VBC-18	72,000	87.1
Burnham	LE1-	74,000	86.7
Burnham	LEDV1-	74,000	86.7
New Yorker	MICROTEK3-1	74,000	86.7
New Yorker	MICROTEKDV-1	74,000	86.7

MOST EFFICIENT OIL BOILERS - HOT WATER (cont.)

Manufacturer	Model	Btu/hr	AFUE (%)
60,000 - 76,000 Btu/hr heating capacity (cont.)			
Buderus	G115-21	74,000	86.3
Burnham	V73WR	73,000	86.2
77,000 - 93,000 Btu/hr heating capacity			
Axeman-Anderson	OL-91D	80,000	88.0
Crown	CTPR-3	92,000	87.6
Dunkirk	DPFO-4T-.65	80,000	87.5
Melvin	CSTO-4T-.65	80,000	87.5
Ultimate	PFO-4T-.65	80,000	87.5
Dunkirk	DPFO-5-.75	93,000	87.5
Dunkirk	PFO-5-.75	93,000	87.5
Melvin	CSTO-5-.75	93,000	87.5
Ultimate	PFO-5-.75	93,000	87.5
Viessmann	VBC-22	92,000	87.2
Axeman-Anderson	OL-105D	92,000	87.2
Thermo-Dynamics	BY-75D	91,000	87.0
Trianco-Heatmaker	DMAX 75	91,000	87.0
Trianco-Heatmaker	MAX 75	91,000	87.0
Crown	CTPB-3	92,000	86.6
Peerless	EC/ECT-03-075W	92,000	86.6
Axeman-Anderson	OL-91	80,000	86.4
Weil-McLain	WGO-2	86,000	86.4
94,000 - 110,000 Btu/hr heating capacity			
Axeman-Anderson	74NPO-UD	105,000	88.7
Dunkirk	DPFO-5T-.80	99,000	87.5
Energy Kinetics	EK-1	104,000	87.5
Melvin	CSTO-5T-.80	99,000	87.5
Ultimate	PFO-5T-.95	99,000	87.5
Axeman-Anderson	PVT-119HD	102,000	86.7
Axeman-Anderson	74NPO-U	105,000	86.5
Axeman-Anderson	OL-119D	103,000	86.4
Buderus	G115-28	98,000	86.2
Burnham	V74WR	98,000	86.2
Weil-McLain	WGO-3R	98,000	86.2
Weil-McLain	WTGO-3R	98,000	86.2
Burnham	V75WR	110,000	86.2
111,000 - 127,000 Btu/hr heating capacity			
Dunkirk	DPFO-6T-.95	117,000	87.5
Melvin	CSTO-6T-.95	117,000	87.5
Ultimate	PFO-6T-.95	117,000	87.5
Viessmann	VBC-33	116,000	87.2
Thermo-Dynamics	NYS-100	122,000	86.7
Trianco-Heatmaker	DMAX 100	121,000	86.5
Trianco-Heatmaker	MAX 100	121,000	86.5

MOST EFFICIENT OIL BOILERS - HOT WATER (cont.)

Manufacturer	Model	Btu/hr	AFUE (%)
111,000 - 127,000 Btu/hr heating capacity (cont.)			
Carrier	BW*AAH000111ABAA	111,000	86.4
Crown	BD-111	111,000	86.4
Dunkirk	4E.90	111,000	86.4
Lennox	COWB-4-0.90	111,000	86.4
Pennco	4K.90	111,000	86.4
Sears	229.944340	111,000	86.4
Thermo-Dynamics	NYV-100	122,000	86.4
Weil-McLain	WGO-4R	122,000	86.2
Weil-McLain	WTGO-4R	122,000	86.2
128,000 - 150,000 Btu/hr heating capacity			
Dunkirk	DPFO-7-1.05	130,000	87.6
Melvin	CSTO-7-1.05	130,000	87.6
Ultimate	PFO-7-1.05	130,000	87.6
Viessmann	VBC-40	146,000	87.3
Axeman-Anderson	87NPO-UD	134,000	87.2
Axeman-Anderson	74NPOD	133,000	86.6
Thermo-Dynamics	NYS-110	133,000	86.6
Thermo-Dynamics	NYV-110	131,000	86.3
Hydrotherm	PB-150W	133,000	86.2
Smith Cast Iron	8-W-4L	133,000	86.2

MOST EFFICIENT OIL BOILERS - STEAM

Manufacturer	Model	Btu/hr	AFUE (%)
77,000 - 93,000 Btu/hr heating capacity			
Columbia	CSFH-365	79,000	86.0
Utica	SFH365S	78,000	84.0
Hydrotherm	PB-105S	90,000	85.6
Smith Cast Iron	8-S-3L	90,000	85.6
Peerless	EC/ECT-03-075S	91,000	85.4
Thermo-Dynamics	LMD-75	90,300	85.2
Smith Cast Iron	BB-14A-S-3L	89,000	83.7
94,000 - 127,000 Btu/hr heating capacity			
Columbia	CSFH-4100	120,000	86.0
Peerless	EC/ECT-03-100S	120,000	84.4
Burnham	V74SR	96,000	84.0
Burnham	V75SR	108,000	84.0
Thermo-Dynamics	LMD-100	118,300	84.0
Weil-McLain	SGO-3	114,000	83.8
Smith Cast Iron	BB-14A-S-3M	101,000	83.5
Hydrotherm	PB-120S	112,000	83.4
Smith Cast Iron	8-S-3H	112,000	83.4
Peerless	WBV-03-085S	101,000	83.1

MOST EFFICIENT OIL BOILERS - STEAM (cont.)			
Manufacturer	Model	Btu/hr	AFUE (%)
128,000 - 160,000 Btu/hr heating capacity			
Columbia	CSFH-5125	151,000	86.0
Hydrotherm	PB-150S	132,000	85.5
Smith Cast Iron	8-S-4L	132,000	85.5
Peerless	EC/ECT-04-125S	151,000	85.1
Hydrotherm	PB-180S	149,000	84.7
Smith Cast Iron	8-S-4H	149,000	84.7
Slant/Fin	L30PZ	134,000	84.1
Burnham	V-76SR	138,000	84.0
Weil-McLain	SGO-4	144,000	84.0
Smith Cast Iron	BB-14A-S-4L	144,000	83.3
Peerless	WBV-04-125S	149,000	83.0

MOST EFFICIENT HEAT PUMPS

Heat pumps are far more energy efficient than electric resistance heat, including electric furnaces, and they can be used both for heating and air conditioning. But before deciding to replace your present system with a conventional heat pump, you should carefully look into whether they make sense in your climate. Air-source heat pumps rely on the outside air as the heat source in the wintertime. The colder that air, the worse the energy performance (the lower the COP). In fact, if the outside temperature falls below about 20°F or 30°F, most air-source heat pumps become no more efficient than electric resistance heating. For this reason, air-source heat pumps are much more common in warmer climates, where winter temperatures seldom fall below 30°F and where summer cooling loads are considerable. On the other hand, geothermal (or ground-source) heat pumps use the ground as a source of heat in the winter and also as a sink for heat in the summer, and they are more efficient than conventional, air-source heat pumps. Geothermal heat pumps are described further in their own section below.

Heat pumps are rated for heating and cooling—both in terms of capacity and efficiency. Capacity ratings are generally in Btu/ hr, as they are with other heating systems. Heating efficiency for air source heat pumps is indicated by the heating season performance factor, or HSPF, which tells you the ratio of the seasonal heating output in Btu divided by the seasonal power consumption in watts. The seasonal energy efficiency rating, or SEER, tells you the seasonal cooling performance. Cooling will be addressed in Chapter 5.

The most efficient air-source heat pumps on the market are listed in the following charts. These range from HSPF 7.8 to 9.7, with smaller capacity heat pumps tending to be at the lower end of the efficiency range. To qualify for an ENERGY STAR® label, an air-source heat pump must have an HSPF of 7 or higher and a SEER of 12 or higher. There are many more Energy Star heat pumps than the models we list; the listings below highlight only the very most efficient.

			Btu/hr Cool	SEER	Btu/hr Heat	HSPF
MOST EFFICIENT CONVENTIONAL (AIR-SOURCE) HEAT PUMPS						
Brand	Outdoor Unit	Indoor Unit	Btu/hr Cool	SEER	Btu/hr Heat	HSPF
Capacity: Approximately 1.5 tons						
Bryant	697CN018-B	FK4CNF002+LLS	19,000	15.00	16,600	8.50
Carrier	38YRA01831	40FKA/FK4CNF002+ LLS	19,000	15.00	16,600	8.50
Rheem/Ruud	*PPA-019JA	*BHJ-17+RCHJ-24A1	19,000	14.00	18,000	8.50
Weatherking	WPPA-019JA	WBHJ-17+RCHJ-24A1	19,000	14.00	18,000	8.50
Rheem/Ruud	*PNJ-019JA	*BHJ-17+RCHJ-24A1	19,000	13.50	18,000	8.50
Weatherking	WPNJ-019JA	WBHJ-17+RCHJ-24A1	19,000	13.50	18,000	8.50
Amana	RHE18A2A	CHA30TCC+BBC36A2A	20,200	14.10	18,200	8.20
Rheem/Ruud	*PPA-018JA	*BHJ-17+RCHJ-24A1	19,000	14.00	18,000	8.00
Weatherking	WPPA-018J	WBHJ-17+RCHJ-24A1	19,000	14.00	18,000	8.00
Trane	TWY018B	TWE031E13	17,300	15.10	16,300	7.80
Capacity: Approximately 2 tons						
Fraser-Johnston/ Luxaire	EABE-F024	GBFD046S17+ NAVSB12+1TV0701	25,400	16.00	23,600	9.25
York	E1RE024S06	G2FD046S17+ N1VSB12+1TV0701	25,400	16.00	23,600	9.25
Coleman Evcon	FRHS0241CD	G*FD046S17+ N*VSB12+TXV	24,800	16.00	23,200	9.00
Airpro/Kenmore	FRHS0241CD	F*FP040N+TXV	24,200	14.50	23,600	9.00
Coleman Evcon/ Guardian	FRHS0241CD	F*FP040N+TXV	24,200	14.50	23,600	9.00
Fraser-Johnston/ Luxaire	EABE-F024	FBFP040+1TV0701	24,200	14.50	23,600	9.00
York	E1RE024S06	F2FP040+1TV0701	24,200	14.50	23,600	9.00
Rheem/Ruud	*PPA-025JA	*BHJ-17+RCHJ-24A2	23,600	14.00	23,200	8.70
Weatherking	WPPA-025JA	WBHJ-17+RCHJ-24A2	23,600	14.00	23,200	8.70
Lennox	HP27-024-1P	CB31MV-41-1P	25,800	15.05	24,200	8.60
Carrier	38YSA02431	40FK/FKCNF002	23,600	14.50	24,000	8.60
American Standard	6H4024A+ BAYINS001	TWE040E13	25,600	14.55	22,800	8.50
Trane	TWY024B	TWE037E13	24,800	14.80	21,800	8.10
Capacity: Approximately 2.5 tons						
Coleman Evcon	FRHS0301CD	G*FD048S21+ N*VSC16+TXV	31,000	15.00	30,000	9.20

MOST EFFICIENT CONVENTIONAL (AIR-SOURCE) HEAT PUMPS (cont.)

Brand	Outdoor Unit	Indoor Unit	Btu/hr Cool	SEER	Btu/hr Heat	HSPF
Capacity: Approximately 2.5 tons (cont.)						
Fraser-Johnston/						
Luxaire	EABE-F030	GBFD048S21+	31,000	15.00	30,000	9.20
		NAVSC16+1TV0702				
York	E1RE030S06	G2FD048S21+	31,000	15.00	30,000	9.20
		N1VSC16+1TV0702				
Carrier	38YSA03031	40FKA/FK4CNF005	30,000	15.50	32,000	9.00
Lennox	HP27-030-1P	CB31MV-41-1P	28,400	15.05	26,400	9.00
Trane	TWY030B	TWE040E13	31,600	14.30	29,600	9.00
Lennox	HP26-030-7P	CB31MV-41-1P	31,400	14.70	30,600	8.85
Coleman Evcon	FRHS0301CD	G*FD046S17+	30,400	15.00	30,400	8.50
		N*VSB12+TXV				
Fraser-Johnston/						
Luxaire	EABE-F030	GBFD046S17+	30,400	15.00	30,400	8.50
		NAVSB12+1TV0702				
York	E1RE030S06	G2FD046S17+	30,400	15.00	30,400	8.50
		N1VSB12+1TV0702				
Capacity: Approximately 3 tons						
Trane	TWY036B	TWE040E13	36,800	14.60	35,000	9.70
Carrier	38YSA03631	40FKA/FK4CNB006	35,400	14.70	35,400	9.40
American Standard	6H4036A	TWE040E13	40,000	14.30	36,200	9.25
Coleman Evcon	FRHS0361CD	G*FD060S24+	37,000	15.60	37,400	9.00
		N*VSD20+TXV				
Fraser-Johnston/						
Luxaire	EABE-F036	GBFD060S24+	37,000	15.60	37,400	9.00
		NAVSD20+1TV0702				
York	E1RE036S06	G2FD060S24+	37,000	15.60	37,400	9.00
		N1VSD20+1TV0702				
Coleman Evcon	FRHS0361CD	G*FD048S21+	36,400	15.00	37,400	9.00
		N*VSC16+TXV				
Fraser-Johnston/						
Luxaire	EABE-F036	GBFD048S21+	36,400	15.00	37,400	9.00
		NAVSC16+1TV0702				
York	E1RE036S06	G2FD048S21+	36,400	15.00	37,400	9.00
		NVSC16+1TV0702				
Carrier	38YDA03630	40FKA/FK4CNB006	40,000	16.10	35,000	8.80
Lennox	HP27-036-1P	CB31MV-51-1P	35,000	15.00	32,200	8.80
Trane	TWZ036A	TWE040E13	37,600	17.65	33,000	8.50
American Standard	6H6036A	ADD120R9V5+	37,200	17.20	32,400	8.50
		TXH054A4+TAYTXV-3				
Trane	TWZ036A	TDD100R9V5+	36,400	17.00	31,600	8.40
		TXC037S3				
Capacity: Approximately 3.5 tons						
Trane	TWY042B	TWE065E13	43,500	14.40	39,500	9.40

MOST EFFICIENT CONVENTIONAL (AIR-SOURCE) HEAT PUMPS (cont.)						
Brand	Outdoor Unit	Indoor Unit	Btu/hr Cool	SEER	Btu/hr Heat	HSPF
Capacity: Approximately 3.5 tons (cont.)						
Coleman Evcon	FRHS0421CD	G*FD061H24+ N*VSD20+TXV	44,500	15.00	43,000	9.25
Airpro/Kenmore	FRHS0361CD	F*FP060N+TXV	44,000	15.20	42,500	9.20
Coleman Evcon/ Guardian	FRHS0421CD	F*FV060N+TXV	44,000	15.20	42,500	9.20
Fraser-Johnston/ Luxaire	EABE-F042	GBFD060S24+ NAVSD20+1TV0703	44,000	15.20	42,500	9.20
Fraser-Johnston/ Luxaire	EABE-F042	FBFV060+1TV0702	44,000	15.20	42,500	9.20
York	E1RE042S06	F2FV060+1TV0702	44,000	15.20	42,500	9.20
York	E1RE042S06	G2FD060S24+ N1VSD20+1TV0703	44,000	15.20	42,500	9.20
Coleman Evcon	FRHS0421CD	G*FD048S21+ N*VSC16+TXV	43,000	14.35	43,000	9.10
Fraser-Johnston/ Luxaire	EABE-F042	GBFD048S21+ NAVSC16+1TV0703	43,000	14.35	43,000	9.10
York	E1RE042S06	G2FD048S21+ N1VSC16+1TV0703	43,000	14.35	43,000	9.10
Carrier	39YSA04231	40FKA/FK4CNB006	41,000	14.20	41,500	8.80
Lennox	HP21-511 -4P,5P	CB31MV-**-1P	44,000	16.15	40,000	8.50
American Standard	6H6048A	TWE040E13	46,000	15.55	41,000	8.00
Capacity: Approximately 4 tons						
Airpro/Kenmore	FRHS0421CD	G*FD061H24+ N*VSD20+TXV	47,000	15.30	44,000	9.20
Coleman Evcon	FRHS0481BD	G*FD061H24+ N*VSD20+TXV	47,000	15.30	44,000	9.20
Fraser-Johnston/ Luxaire	EABE-F048	GBFD061H24+ NAVSD20+1TV0703	47,000	15.30	44,000	9.20
York	E1RE048S06	G2FD061H24+ N1VSD20+1TV0703	47,000	15.30	44,000	9.20
Coleman Evcon/ Guardian	FRHS0481BD	F*FV060N+TXV	46,500	15.00	44,500	9.00
Fraser-Johnston/ Luxaire	EABE-F048	FBFV060+1TV0703	46,500	15.00	44,500	9.00
York	E1RE048S06	F2FV060+1TV0703	46,500	15.00	44,500	9.00
Coleman Evcon	FRHS0481BD	G*FD060S24+ N*VSD20+TXV	46,500	15.10	44,000	8.75
Fraser-Johnston/ Luxaire	EABE-F048	GBFD060S24+ NAVSD20+1TV0703	46,500	15.10	44,000	8.75

			Btu/hr		Btu/hr	
MOST EFFICIENT CONVENTIONAL (AIR-SOURCE) HEAT PUMPS (cont.)						
Brand	Outdoor Unit	Indoor Unit	Cool	SEER	Heat	HSPF
Capacity: Approximately 4 tons (cont.)						
York	E1RE048S06	G2FD060S24+ N1VSD20+1TV0703	46,500	15.10	44,000	8.75
American Standard	6H6048A	TWE065E13	48,500	16.05	43,000	8.25
Trane	TWZ048A	TWE065E13	47,500	15.50	43,000	8.25

GROUND-SOURCE HEAT PUMPS OR "GEOEXCHANGE"

Ground-source (or geothermal) heat pumps (referred to as "GeoExchange" units) use the earth or groundwater as the heat source instead of outside air. Because temperatures underground are nearly constant year-round—warmer than the outside air during the winter and cooler than the outside air during the summer—a GeoExchange unit will be more efficient than an air-source unit. The stable underground temperature allows geothermal systems to be rated for heating efficiency in terms of coefficient of performance, COP, and cooling efficiency in terms of energy efficiency ratio, EER. As shown in the following tables, GeoExchange heat pumps are easily 25%–45% more efficient than new air-source heat pumps.

GeoExchange heat pumps are less common than air-source heat pumps, and are often more expensive and difficult to install. But the dramatic improvement in efficiency yields attractive lifecycle cost savings through low energy bills, compared to conventional heat pumps. In addition, since the heat exchange pipes are underground and protected from the elements, GeoExchange units require less maintenance than conventional heat pumps, and there is no outside noise typically associated with conventional heat pumps and air conditioners. As an added bonus, GeoExchange heat pumps can provide inexpensive water heating during much of the year.

GeoExchange pipes are buried in the ground, usually in long, shallow (3'–6' deep) trenches or in one or more vertical holes. The particular method used will depend on the experience of the installer, the size of your lot, the subsoil, and the landscape. Of course, it can be counterproductive to damage beautiful shade trees by plowing through their roots for your new heating and cooling system. An experienced loop installer will be able to design a pipe system that is appropriate to your lot and landscape.

Several utilities, in cooperation with government agencies and other groups, have begun marketing GeoExchange systems as a superior alternative to air-source heat pumps, especially for new housing, where the cost and disruption of burying heat exchange coils is not as great as in replacement installations. New construction also provides the opportunity to combine GeoExchange heat pumps with other efficiency measures, such as extra wall and ceiling insulation, and high-efficiency windows. A more efficient building shell allows for a smaller and less costly GeoExchange system, and can lead to home energy bill reductions of 75% compared to conventional new housing using air-source heat pumps.

Utilities and the GeoExchange industry have been very active in the past few years to train installers, reduce system costs, and further improve system performance, reliability, and efficiencies. For more information on GeoExchange heat pumps, including manufacturers, installation, and related issues, please contact the following groups:

International Ground Source
  Heat Pump Association
490 Cordell South
Oklahoma State University
Stillwater, OK 74078-8018
(800) 626-4747
(405) 744-5175
www.igshpa.okstate.edu

Geothermal Heat Pump
  Consortium
701 Pennsylvania Avenue NW
Washington, DC 20004-2696
(888) 255-4436
(202) 508-5500
www.ghpc.org

MOST EFFICIENT GROUND-SOURCE HEAT PUMPS

Manufacturer	Model	Btu/hr Cool	EER	Btu/hr Heat	COP
Capacity: About 2.5 tons cooling; 20,000-24,000 Btu/hr heating					
Waterfurnace	SX*030**	31,200	14.3	21,600	3.5
Bard	GSVS301-A	27,400	16.0	20,400	3.4
Econar	G*23*	24,800	14.5	21,200	3.4
Climate Master	GSV/GSH 030	32,000	16.1	23,000	3.3
Tetco	ESII2.5	26,200	13.8	21,200	3.3
Florida Heat Pump	GT030-*VT/HZ/CF	31,200	15.0	20,600	3.2
Climate Master	GSS30	30,000	14.6	22,000	3.2
Climate Master	PDW30	30,000	14.6	22,000	3.2
Command-Aire	GSUF0301B	30,200	17.1	22,000	3.1
Trane	GSUF0301B	30,200	17.1	22,000	3.1
Capacity: About 3 tons cooling; 24,400-31,900 Btu/hr heating					
Carrier	50YBV042	42,000	17.0	30,000	3.8
Climate Master	VT 042	42,000	17.0	30,000	3.8
Waterfurnace	AT*040**	41,500	17.0	29,200	3.4
Bard	GSVS361-A	34,800	16.0	24,400	3.4
Carrier	50YBV036	34,500	15.5	26,000	3.4
Climate Master	VT 036	34,500	15.5	26,000	3.4

MOST EFFICIENT GROUND-SOURCE HEAT PUMPS (cont.)

Manufacturer	Model	Btu/hr Cool	EER	Btu/hr Heat	COP
Capacity: About 3 tons cooling; 24,400-31,900 Btu/hr heating (cont.)					
Climate Master	PDW 36	36,000	15.0	29,000	3.4
Waterfurnace	AT*034**	34,400	16.0	25,000	3.3
Command-Aire	GSSD 036	36,400	15.7	26,000	3.3
Trane	GSSD 036	36,400	15.7	26,000	3.3
Command-Aire	GSSD 042	41,000	14.3	30,200	3.3
Trane	GSSD 042	41,000	14.3	30,200	3.3
Econar	G*29*	28,600	13.7	25,800	3.3
Capacity: About 4 tons cooling; 32,000-37,900 Btu/hr heating					
Carrier	50YBV048	46,300	16.6	33,200	3.7
Climate Master	VT 048	46,300	16.6	33,200	3.7
Florida Heat Pump	GO048-1VT/HZ/CF	49,000	15.2	34,000	3.4
Climate Master	GSS 42	43,000	15.0	33,000	3.4
Climate Master	PDW 42	43,000	15.0	33,000	3.4
Waterfurnace	AT*046**	47,000	16.1	34,500	3.3
Climate Master	GSV/GSH 042	42,800	15.2	32,800	3.3
Climate Master	GSV/GSH 048	48,700	15.0	37,100	3.3
Command-Aire	GSUG0481	49,000	14.9	33,000	3.3
Trane	GSUG0481	49,000	14.9	33,000	3.3
Command-Aire	GSSD 048	48,500	14.9	37,200	3.3
Trane	GSSD 048	48,500	14.9	37,200	3.3
Capacity: About 4.5 tons cooling; 38,000-42,500 Btu/hr heating					
Carrier	50YBV060	57,500	14.3	38,900	3.4
Climate Master	VT 060	57,500	14.3	38,900	3.4
Climate Master	GSV/GSH 060	61,100	14.1	49,700	3.2
Econar	Q14KW4T	47,500	14.0	38,400	3.2
Climate Master	VP/HP/DP048	48,000	13.5	38,000	3.2
Mammoth	D/F054VLE/HLE/DLE	54,000	13.7	42,000	3.1
Mammoth	G054VLE/HLE/DLE	54,500	13.7	42,500	3.1
Mammoth	D/F054ULE	53,500	13.7	41,500	3.1
Mammoth	G054ULE	54,500	13.7	42,500	3.1
Tetco	ESII4.0	43,000	13.6	39,500	3.1
Capacity: About 5 tons cooling; 43,000-53,000 Btu/hr heating					
Millbrook	CS-HA57	60,200	13.0	51,200	3.3
Waterfurnace	AT*056**	58,000	15.3	44,500	3.2
Waterfurnace	RTV056*1	57,000	14.8	44,500	3.2
Waterfurnace	RTV066*1	65,000	13.8	51,800	3.2
Carrier	50YBV072	70,000	13.7	50,000	3.2
Climate Master	VT 072	70,000	13.7	50,000	3.2
Waterfurnace	SX*058**	56,500	13.4	45,500	3.1
Command-Aire	GSSD 060	59,000	13.3	46,900	3.1
Trane	GSSD 060	59,000	13.3	46,900	3.1
Command-Aire	GSUF0601B	62,500	15.1	45,000	3.0
Trane	GSUF0601B	62,500	15.1	45,000	3.0

MOST EFFICIENT GROUND WATER-SOURCE HEAT PUMPS

Manufacturer	Model	Btu/hr Cool	EER	Btu/hr Heat	COP
Capacity: About 2.0 tons cooling; 24,000-28,900 Btu/hr heating					
Waterfurnace	VLV/VXV024**	23,800	14.2	28,200	4.7
Climate Master	VS/VL/HS/HL 024 *,*	24,500	13.4	28,900	4.6
Carrier	50HQA,HQD,VQA,VQD024 *,*	24,500	13.4	28,900	4.6
Mammoth	*020HHF,VHF	19,500	13.2	26,500	4.6
Tetco	ESII 2.0	21,400	14.2	24,000	4.5
Trane	WPHF021*	21,200	14.9	23,600	4.4
Cold Flow	CF024*	23,800	14.0	28,800	4.4
Command-Aire	GS**0181B	22,600	16.4	25,200	3.9
Trane	GS**0181B	22,600	16.4	25,200	3.9
Command-Aire	GSUF0241B	25,800	14.9	28,600	3.9
Trane	GSUF0241B	25,800	14.9	28,600	3.9
Capacity: About 2.0-2.5 tons cooling; 29,000-32,900 Btu/hr heating					
Waterfurnace	SX*024**	25,500	15.1	30,700	5.3
Climate Master	GSV/GSH 024	24,900	15.0	31,100	4.8
Florida Heat Pump	EM024-*VT/HZ/CF	24,000	14.0	30,000	4.8
Heat Controller	WSH024-*A*	24,000	14.0	30,000	4.8
Addison	WSH024-*A*	24,000	14.0	30,000	4.8
Florida Heat Pump	EM024-*CS	23,600	13.8	29,500	4.7
McQuay	W-CDD/CDE/CMG-1-024-*	25,400	13.8	29,800	4.1
McQuay	CCW024E-*	25,400	13.8	29,800	4.1
Trane	GSSD024	24,200	14.8	29,200	4.0
Command-Aire	GSSD024	24,200	14.8	29,200	4.0
Florida Heat Pump	GT030-*CS	30,800	14.1	32,200	3.9
Command-Aire	GSHC0241B	26,600	16.6	29,800	3.7
Trane	GSHC0241B	26,600	16.6	29,800	3.7
Capacity: About 2.5-3.0 tons cooling; 33,000-38,500 Btu/hr heating					
Waterfurnace	SX*030**	30,600	14.4	36,600	5.0
Climate Master	GSV/GSH 030	29,800	15.1	36,100	4.8
Mammoth	*027HHF,VHF	25,000	13.5	33,000	4.8
Heat Controller	WSH029-*A*	28,300	14.0	34,900	4.7
Addison	WSH029*A*	28,300	14.0	34,900	4.7
Florida Heat Pump	EM028-*VT/HZ/CF	28,000	14.2	33,400	4.5
Tetco	ESII 2.5	27,200	14.0	33,000	4.5
Waterfurnace	VLV/VXV030**	29,500	14.0	35,200	4.5
Florida Heat Pump	EM028-*CS	27,500	14.0	33,000	4.4
Cold Flow	CF031*	30,800	14.0	36,000	4.4
Command-Aire	GSSD030	30,000	14.7	35,400	4.0
Trane	GSSD030	30,000	14.7	35,400	4.0
DeMarco	GT030-*VT,HZ,CF	31,400	14.2	33,000	4.0
Florida Heat Pump	GT030-*VT,HZ,CF	31,400	14.2	33,000	4.0
Command-Aire	GSUF0301B	32,000	16.6	33,800	3.9
Trane	GSUF0301B	32,000	16.6	33,800	3.9
Capacity: About 3.0-3.5 tons cooling; 39,000-42,500 Btu/hr heating					
Florida Heat Pump	EM036-*VT/HZ/CF	35,000	14.0	42,000	4.5

MOST EFFICIENT GROUND WATER-SOURCE HEAT PUMPS (Cont.)

Manufacturer	Model	Btu/hr Cool	EER	Btu/hr Heat	COP
Capacity: About 3.0-3.5 tons cooling; 39,000-42,500 Btu/hr heating (cont.)					
Florida Heat Pump	EM031-*VT/HZ/CF	32,000	14.0	42,000	4.5
Addison	WSH034-*A*	33,500	13.6	42,500	4.5
Heat Controller	WSH034-*A*	33,500	13.6	42,500	4.5
Tetco	ESII 3.0	35,200	13.8	40,500	4.4
Florida Heat Pump	EM031-*CS	31,400	13.8	41,000	4.4
Florida Heat Pump	EM036-*CS	34,500	13.8	41,000	4.4
Trane	GSHC0361B	36,600	15.2	42,000	4.3
Command-Aire	GSHC0361B	36,600	15.2	42,000	4.3
Trane	GSSD036	37,000	15.0	40,500	4.2
Command-Aire	GSSD036	37,000	15.0	40,500	4.2
Trane	GSUF0361B	36,800	16.2	40,000	4.0
Command-Aire	GSUF0361B	36,800	16.2	40,000	4.0
Florida Heat Pump	GT036-*VT,HZ,CF	36,000	13.9	39,000	3.8
DeMarco	GT036-*VT,HZ,CF	36,000	13.9	39,000	3.8
Capacity: About 3.5-4.0 tons cooling; 43,000 -50.000 Btu/hr heating					
Climate Master	GSV/GSH 036	34,600	15.0	42,900	4.7
Waterfurnace	SX*036**	36,000	13.6	44,500	4.7
Mammoth	*035HHE,VHE	37,000	14.1	45,000	4.6
Tetco	ESII 3.5	41,000	13.7	46,000	4.4
Cold Flow	CF036*	35,400	13.7	43,600	4.4
Trane	WPHF0404	39,500	14.8	43,000	4.3
Trane	GSSD042	43,000	14.9	49,500	4.2
Command-Aire	GSSD042	43,000	14.9	49,500	4.2
Trane	GSUF0421*	43,500	15.2	45,500	4.1
Command-Aire	GSUF0421A	43,500	15.2	45,500	4.1
Command-Aire	GSHC0421B	41,000	14.0	50,500	4.1
Trane	GSHC0421B	41,000	14.0	50,500	4.1
Hydro Delta	03-038-WTARH	36,800	13.5	44,000	4.1
Capacity: About 4.0-5.0 tons cooling; 51,000+ Btu/hr heating					
Waterfurnace	SX*042**	42,000	14.0	54,800	5.1
Climate Master	GSV/GSH 042	42,000	14.6	51,300	4.8
Climate Master	GSV/GSH 048	46,700	14.6	55,800	4.8
Koldwave	6K48CMHT	51,300	15.2	55,000	4.7
Mammoth	*054HHE,VHE	53,500	14.2	67,000	4.6
Cold Flow	CF048V	50,000	14.0	63,000	4.6
Mammoth	*045HHF,VHF	44,500	14.0	58,000	4.6
Waterfurnace	SX*058**	55,000	13.6	69,000	4.5
Florida Heat Pump	GO048-1VT/HZ/CF	49,000	13.8	52,000	4.3
Tetco	ESII 4.0	46,000	13.7	52,000	4.3
Command-Aire	GSSD048	49,000	14.3	56,000	4.2
Trane	GSSD048	49,000	14.3	56,000	4.2
Command-Aire	GSUF0481B	51,000	15.9	54,000	4.1
Trane	GSUF0481B	51,000	15.9	54,000	4.1
Trane	GSUF0601B	64,500	15.3	66,000	4.0
Command-Aire	GSUF0601B	64,500	15.3	66,000	4.0
Trane	WPHF0571	57,500	14.5	60,000	4.0

UPGRADING YOUR EXISTING HEATING SYSTEM

Even if you aren't about to go out and buy a state-of-the-art, high-efficiency heating system, you can still probably realize substantial savings by boosting the efficiency and performance of your present system. Measures described here fall into two categories: maintenance and modifications.

Regular Maintenance and Tune-ups

Proper maintenance can have a big effect on fuel bills and should be performed on a routine basis. Maintenance operations are separated into two groups: those that can be performed by the homeowner or renter, and those that should be performed by a heating system service technician.

Maintenance by the Resident

When performing maintenance on hot water and steam heating systems, you may come in contact with dangerously hot water and steam. Use caution. If you're uncertain about how to do something, call a service technician (or your landlord, if you rent).

■ *Clean air filters.* The filters on warm-air furnaces and heat pumps should be checked once a month during the heating season and cleaned or replaced as necessary. Dust blocks the air flow and forces the blower to work harder, which raises electric bills and can lead to blower failure. Filters cost about $1 apiece and can usually be purchased at hardware stores.

Replace furnace air filter monthly during the heating season.

■ *Clean registers.* Warm-air registers should be kept clean and should not be blocked by furniture, carpets, or drapes.

Clean warm-air registers.

■ *Keep baseboards and radiators clean and unrestricted* by furniture, carpets, or drapes. Air needs to freely circulate through them from underneath. Also, do not cover tops of radiators.

■ *Bleed trapped air from hot water radiators.* Trapped air keeps radiators from performing properly. Use a radiator key to bleed air out of hot-water radiators once or twice a season. Hold a pan under the valve and open it until all the air has escaped and only water comes out. If you are not mechanically inclined, you may want to have the technician show you how to do it the first time.

■ *Follow prescribed maintenance* for steam heat systems, such as maintaining water level, removing sediment, and making sure air vents are working. Check with your heating system technician for specifics on these measures and use caution: steam boilers produce high-temperature steam under pressure.

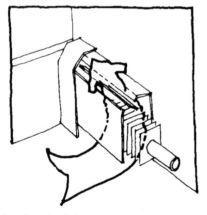
Baseboard radiator.

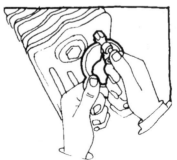
Bleeding hot-water radiator.

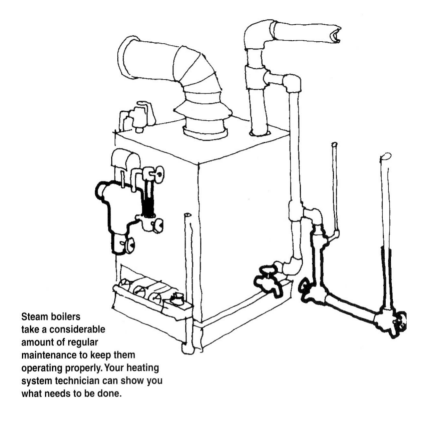

Steam boilers take a considerable amount of regular maintenance to keep them operating properly. Your heating system technician can show you what needs to be done.

Maintenance and Tune-ups by Heating System Technicians

Oil-fired systems should be tuned up and cleaned every year, gas-fired systems every two years, and heat pumps every two or three years. Regular tune-ups not only cut heating costs, but they also increase the lifetime of the system, reduce breakdowns and repair costs, and cut the amount of carbon monoxide, smoke, and other pollutants pumped into the atmosphere. System modifications to improve energy efficiency can be made when the service technician is there for the annual tune-up (see the next section).

The company that sells oil or gas usually has trained technicians who can test your furnace or boiler, clean it, and tune it for optimum efficiency. Independent contractors provide this service as well. A complete tune-up usually costs $50–100 and reduces your heating bill from 3–10%. Some companies perform these services as part of a regular service contract. Check to make sure that all of the services listed below are included.

Tests the Technician Should Perform During a Tune-up. The service technician will perform a number of tests to determine the system's performance and efficiency. Incomplete combustion of fuel and excessively high flue gas temperatures are the two main contributors to low efficiency. If the technician cannot get the combustion efficiency up to at least 75% after tuning it up, you should consider installing a new system or at least modifying your present system to increase its efficiency. Note that the combustion efficiency is different from annual fuel utilization efficiency (AFUE). For older burners, the AFUE can be estimated by multiplying the combustion efficiency by 0.85. Thus, if the combustion efficiency is 75%, the AFUE is around $0.75 \times 0.85 = 64\%$.

The technician should measure the efficiency of your system both before and after tuning it up and provide you with a copy of the results. Combustion efficiency is determined indirectly, based on some of the following tests: 1) flue temperature; 2) percent carbon dioxide or oxygen; 3) smoke number; 4) carbon monoxide; and 5) draft. These are discussed below:

■ *Flue temperature.* High flue gas temperatures mean that a lot of heat (and money) is being lost up the chimney. Typical flue temperatures are:

Gas	300–600°F
Gas (condensing system)	100–200°F
Oil	400–600°F
Oil (flame retention burner)	300–500°F

■ *Carbon dioxide.* Carbon dioxide is the primary end product of fossil fuel combustion. Too little carbon dioxide indicates incomplete combustion. For an oil burner, the CO_2 concentration should measure between 10 and 12%. For gas, it should be between 7 and 9%. If an oxygen reading is taken instead, it should be between 3 and 6% for oil systems, or between 5 and 7% for gas systems.

■ *Smoke.* Smoke indicates lack of complete combustion and is usually not present in gas systems. On a scale of 0 to 10, the smoke number should be no higher than 1.

■ *Carbon monoxide (gas only).* Carbon monoxide indicates incomplete combustion and should be kept below one-tenth part per million (0.1 ppm) for safety reasons.

■ *Draft.* Correct draft promotes complete combustion and reduces net loss up the chimney. A pressure gauge measures the overfire draft through the combustion chamber, and the breach draft through the flue pipe. Overfire draft pressure should be between 0.01 and 0.02 inches of water, and breach draft should be between 0.02 and 0.04 inches higher than the overfire draft. If you have a sealed combustion or induced draft system (a fan pushing exhaust gases out), this is less important.

Cleaning. Parts to be cleaned include the burner (nozzle, electrodes, filters), combustion chamber, heat exchanger surfaces, oil line filter, and flue pipe. (Oil nozzles and filters are often replaced rather than cleaned.) Sediment should be removed from the boiler and steam lines; corrosion inhibitors may be added to the boiler.

Adjustments. Air and fuel flow adjustments will be made based on the results of the efficiency testing. The internal thermostat on the furnace or boiler (aquastat or fan thermostat) should be calibrated to turn on and off at the appropriate temperatures.

Pumps and Fans. Pumps and fans should be inspected and lubricated if necessary.

HEATING SYSTEM MODIFICATIONS

Even if your heating system is fairly efficient, there are probably still some modifications that could be made to further increase efficiency. Often, it makes more economic sense to modify your existing system than to replace it, at least if it's in pretty good shape and less than 15 or 20 years old. A service technician or energy auditor should be able to help you decide which approach makes the most sense. Common heating system modifications are described in the following few pages, divided into those that can be made by the homeowner and those that should be done by a technician.

Modifications You Can Do Yourself

All of these modifications deal with heat distribution—getting the most heat from your furnace or boiler to the rooms in your house. These should be done regardless of the tested combustion efficiency of the heating system.

■ *Pipe insulation.* All hot-water and steam pipes passing through unheated areas should be wrapped with insulation. Use specially made foam or fiberglass pipe insulation, which can cost $0.30–0.80 per foot and saves about $0.50 per foot per year. Try to use insulation with a wall thickness of at least ¾" for fiberglass, and ½" for foam. Do not wrap steam pipes with foam as it could melt. Most steam pipes will be wrapped already with asbestos, which should not present a health hazard as long as it is well-sealed, not flaky, and not in a living space. If some of the white protective sheathing is missing, contact an asbestos abatement contractor.

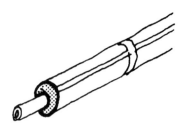

Insulate hot-water pipes that run though unheated spaces. Be sure to use insulation that can withstand high temperatures.

■ *Duct Insulation.* First, seal all joints and seams with quality duct tape (one that complies with UL-181) or mastic (a special paste) to keep hot air from leaking out of the ducts. Duct tape is that strong, silver or grey tape most of us know for its other applications, such as repairing broom handles and car parts. Unfortunately, duct tape can dry up and lose its adhesion over time, especially in unheated spaces. Mastic, on the other hand, spreads easily and dries permanently, and is the preferred material for sealing joints and seams in metal ductwork. (See section on **Modifications by Heating System Technicians** below for more on duct sealing).

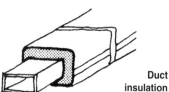

Duct insulation is very important for ducts that pass through unheated spaces.

Secondly, all hot-air ducts passing through unheated spaces should be wrapped with insulation. You can use standard foil-faced fiberglass insulation, keeping the foil facing out and visible; vinyl-faced insulation made especially for ducts; or rigid foam insulation. At least R-5 is recommended in cold climates; R-8 is preferred. Then, seal all joints or seams in the insulation. Use duct tape with standard fiberglass batts and rigid foam; with vinyl-faced duct insulation, use duct tape or double-over and staple the seams.

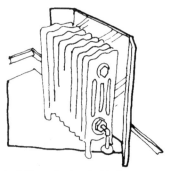

Installing reflectors behind your radiators can improve heat transfer into the room.

■ *Radiator reflectors.* Radiators are designed to heat the living space, but they can lose a lot of heat into the exterior walls they are installed against. You can reduce this loss by placing reflectors between the wall and the radiator. You can make reflectors from foil-covered cardboard, available from many building supply stores. The reflector should be the same size or slightly larger than the radiator. The foil should be periodically cleaned for maximum heat reflection.

Modifications by Heating System Technicians

All these measures for boosting the efficiency of your furnace or boiler require a professional with the proper training and tools.

■ *Reducing system size.* If you have an older gas or oil system, and if you've added insulation, upgraded your windows, or tightened your house, chances are that your burner runs for only a fraction of the time, even in the coldest weather. Having your heating system constantly turn on and off is like driving in stop-and-go traffic: you don't get very good mileage. A simple way to reduce this waste is by decreasing the rate at which oil or gas is fed into the burner. In some cases, however, derating of gas systems violates local building codes and voids manufacturers' warranties. Check with your local code officials or knowledgeable contractors before proceeding.

With oil systems, the service technician can install a smaller nozzle, which costs just a few dollars and can cut fuel bills by 5–10%. Nozzles are sized according to fuel flow rates. The specification plate on your burner should include an acceptable flow range; an average range would be from 0.50 to 1.25 gallons per hour (GPH). Nozzle size should not be re-

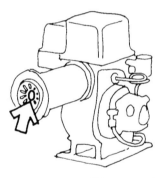

Oil burner nozzle.

duced more than 25–30% below the lowest firing range on the specification plate.

With gas systems, reducing the nozzle (or orifice) size is tricky; costs and savings will vary widely depending on the system. On a unit that is highly oversized, both the nozzle orifice and the baffles could be changed at a cost of $60–80, with possible savings of 10–15%. The orifice size should not be reduced more than 30%. Consult a gas heat service technician to find out if size reduction is possible for your system.

Boilers for steam systems should not be downsized.

■ *Draft reduction (oil only).* The draft test will determine whether excess heat is being lost up the chimney. This problem is particularly common in systems that were converted from coal to oil. If the draft is too high, your service technician should install a barometric damper in the flue pipe. This will cost from $20–80 and can reduce fuel use by 5% or more. If a barometric damper is already there, it may simply need adjustment.

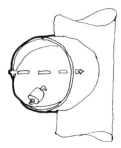

Barometric damper on oil-fired furnace or boiler.

■ *New oil burner installation.* If you have an old, inefficient oil burner but are not ready to replace the whole thing, have a flame-retention burner installed. It will mix oil and air more thoroughly, operate with less air flow, and send less heat up the chimney. In addition, a flame-retention burner will block air flow through the burner when the system is not running, reducing heat loss up the chimney. Flame-retention burners cost $400–500, depending on whether a new combustion chamber and controls are needed. A properly sized flame-retention burner with reduced nozzle size should save 10–20%. You'll do even better, though, replacing the whole furnace or boiler with a state-of-the-art high-efficiency model.

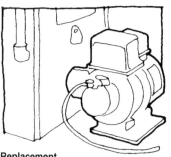

Replacement oil burner.

■ *Modulating aquastats (hot-water boilers only).* An aquastat regulates boiler temperatures, keeping the water within a prescribed temperature range, usually around 180°F. Unfortunately, it will keep the water just as hot even when there is little need for heat, such as during spring and fall months. A modulating aquastat (or outdoor reset) senses outdoor temperature and keeps the boiler water only as hot as needed. Brand-name aquastats such as Enertrol or Mastermind sell for $150–350 and reduce fuel consumption by 5–10%. You can control an aquastat manually as well (see the section below on operating your system).

■ *Time delay relay (hot water boilers only).* Another strategy for controlling boiler water temperature is the time delay relay. When the room thermostat signals a need for heat, water heated earlier is circulated through the radiators without the boiler turning on. If circulation of warm water is not sufficient to heat the home within a specified time, the boiler burner fires to further heat the boiler water. With a time delay relay, circulation of lower temperature boiler water can provide adequate heating during milder weather. A time delay relay can be installed by a contractor for $50-75 and yield savings of 10%.

■ *Pilotless ignition (gas only).* Electronic ignition eliminates the pilot light that ordinarily burns constantly. It costs from $100–250 and has a three- to eight-year payback. Electronic ignition is difficult to install on existing systems and should be undertaken only by someone who is very experienced in this type of work.

■ *Automatic flue damper.* The automatic damper is a metal flap that closes off the flue when the burner shuts off. Flue dampers cost from $125–400 installed and can cut fuel consumption by 3–15%. Savings are highest with steam boilers, large hot-water boilers, and warm-air furnaces that are located in heated spaces, where heated room air can escape up the chimney. If the heating system is located in an unheated basement or if it has a flame retention oil burner, savings will probably be less than 5%. If you have an older oil burner, converting it to a flame retention type is generally a better investment *(see New oil burner installation above).*

There are two types of flue dampers: thermal and electric. Thermal dampers respond to the presence of hot flue gases, while electric dampers are wired directly to the burner. Although electric dampers cost more, they also save more, usually making the extra

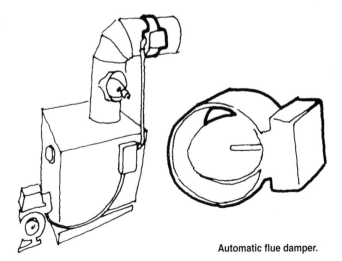

Automatic flue damper.

cost worthwhile. Caution: flue dampers are not suitable for all gas heating systems. Ask your service technician whether a flue damper is appropriate for your system.

■ *Flue economizer.* Flue economizers are devices that recover heat from the hot gases going up the flue. They are expensive, costing $200–800, and generally not recommended. It is almost always better to replace or modify inefficient systems. If you install a flame retention burner instead, for example, you will improve combustion efficiency and reduce the amount of heat going up the chimney rather than trying to recover that heat as an afterthought. Flue economizers also frequently have corrosion problems.

■ *Duct sealing.* In homes heated with warm-air heating, ducts should be inspected and sealed to ensure adequate airflow and eliminate loss of heated air. It is not uncommon for ducts to leak as much as 15-20% of the air passing through them. And, leaky ducts can bring additional dust and humidity into living spaces. Thorough duct sealing costs several hundred dollars but can cut heating and cooling costs in many homes by 20%.

A contractor can test your ducts to determine the extent and location of leaks. Traditionally, ducts are sealed using mastic, which is applied to the outside of duct joints and other leak sites. A new alternative to mastic is aerosol-based duct sealing ("Aeroseal"). A machine connected to the ductwork blows a latex aerosol throughout ducts to seal leaks from the inside. This system can reach leaks in hard to reach or inaccessible spaces and effectively seals leaks up to 1/4" in diameter. Contractors around the country are receiving training in this new technology; see www.aeroseal.com to find franchises in your area.

■ *Adjustable radiator vents and valves.* To reduce heat flow to unused rooms, valves on some hot-water radiators may be turned down. Valves on steam radiators should always be completely on or off, not in-between. An alternative for steam radiators is to install an adjustable air vent, typically costing about $10–15 at hardware and heating supply stores. These vents are screwed onto the radiator in place of existing vents, and they control how much steam gets into the radiator to heat it up.

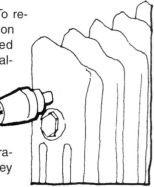

Adjustable radiator valve.

You can get even greater control with either steam or hot- water radiators by installing thermostatic radiator valves. These valves allow you to select the temperature of each room. When the designed temperature is reached, the valve shuts the radiator off. These valves cost $50–125 each installed, and can be a less expensive way to create separate heating zones, compared to repiping the whole house.

■ *Clock thermostats.* Setting the thermostat manually works well but is inconvenient. More convenient is a clock thermostat that will turn on the heat a half-hour before your alarm goes off in the morning. Some clock thermostats have several different set-back periods, helping you save energy when you go off to work and the kids leave for school. Clock thermostats cost from about $40 for simpler models, which allow one or two set-back periods each day, to $150 for models that allow separate programming for each day of the week. Most clock thermostats will pay for themselves in about a year.

Clock thermostat.

■ *Tankless coil water heaters.* If domestic hot water is provided by your heating system boiler with a tankless coil, then during the summer the boiler must operate constantly just to provide hot water for showers, washing dishes, etc. There are several ways to avoid this waste. The simplest but least convenient is to install a timer switch so that you can turn your whole heating system off at night and when you are away during the summer months. Another option is to install a stand-alone, gas-fired water heater to use in the summer when your heating system is off. A third option is to install an indirect water heater that draws heat from the boiler (like a tankless coil), but stores the hot water so that the boiler does not need to run as frequently. This option is usually the most cost-effective alternative to tankless coils in cold climates.

A fourth solution may be to buy a solar water heater. In most areas of the country, solar water heaters can provide nearly 100% of summertime hot water needs, thus complementing your wintertime boiler-produced hot water perfectly. Although this system requires a large capital investment, it is the best solution from an environmental standpoint, because solar energy produces virtually no pollutants, in contrast to fossil fuels and electricity. (See Chapter 6 for more information on water heating.)

OPERATING YOUR HEATING SYSTEM
FOR MAXIMUM EFFICIENCY

How you operate the controls of your heating system can have a large effect on your heating bills. People living in identical houses can have utility bills that vary widely, with some families paying 50% or even up to 100% more than others.

■ *Thermostats.* Turn down thermostats in unused rooms and when you don't need the heat. In most homes, you can save about 2% of your heating bill for each degree that you lower the thermostat. Turning down the thermostat from 70°F to 65°F, for example, saves about 10% ($100 saved per $1,000 of heating cost). Setting your thermostat back 10°F for eight hours at night can save about 7% ($70 saved per $1,000 of heating cost). Clock thermostats, which automatically adjust the temperature setting one or more times per day, are widely available (see previous section on modifications).

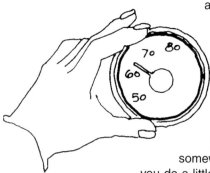

You might also be able to turn the thermostats down somewhat when you're in the room if you do a little buttoning up of the house (see Chapter 2). Turning down thermostats even when you are in a room doesn't mean being uncomfortable. In fact, you can actually be more comfortable at a lower temperature setting under the right conditions. Eliminating temperature stratification in a room where the floor is a lot colder than the ceiling will help the most, and getting rid of air infiltration and cold drafts is the best way to do it. Covering windows at night with blinds or drapes also helps, as does higher humidity. Buy some house plants or a humidifier, but don't add so much moisture that you start seeing condensation on your windows.

Setting the thermostat back at night will save a lot of energy and money.

■ *Aquastat (hot-water boilers).* This is the thermostat that regulates the temperature of the hot-water boiler. Normally, the aquastat keeps water in the boiler around 160–180°F. In milder weather, however, you don't need boiler water that hot. The aquastat can be set manually to 120–140°F, reducing fuel consumption by 5–10%. (If your boiler has a tankless coil for domestic hot water, you may not be able to turn the aquastat down this far.) The aquastat control is usually located in a metal box connected to the boiler. If you cannot locate it, ask your service technician for assistance. See the modifications section for information about modulating aquastats.

■ *Fan thermostat (warm-air furnaces).* This controls the fan that blows warm air through the ducts. If it is set too high, the fan comes on too late and shuts off too soon, leaving heat in the furnace to escape up the flue. If you can't find the fan thermostat, have the technician point it out. If it has two settings, set it to come on at 100-110°F and to shut off at 80-90°F. If it has one setting, set it at 100-110°F. Thermostats with a single setpoint may have a variable deadband which determines the maximum temperature fluctuation from the setpoint. In this case, the variable deadband should be set to 20°F. If you feel cold air blowing out of your registers at the end of a heating cycle, the shut-off temperature is set too low. Raise it enough to eliminate those cold drafts. Since common practice is to set the fan thermostat too high, adjustment can reduce energy use by 5-10% in most homes.

■ *Duct dampers (warm-air furnaces).* Often, air ducts have dampers (adjustable metal flaps) in them to control flow. Shut off or turn down dampers and registers that heat the basement. Other dampers and registers can be turned down or off to control heat flow to various rooms. Unused rooms should be kept cooler than occupied rooms to save energy. Don't close off too much airflow, though, because it may cause trouble for the fan.

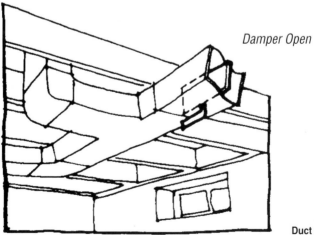

Damper Open

Duct damper.

RECOMMENDATIONS

1. Try to button up your house as much as possible before installing new heating equipment, because you might be able to buy a smaller, less expensive system. Energy conservation is cheaper than burning expensive fossil fuels. Have an energy auditor help you to decide which measures are cost-effective. The auditor can also help you size a new heating system properly. But even if you aren't going to insulate further or weatherstrip, go ahead with the heating system replacement or improvements.

2. Electric resistance heating is not recommended because electricity is much more expensive than gas or oil. Electric heat pumps may be appropriate in your area; conventional air-source heat pumps can be economical in areas with mild winters and substantial cooling loads, but they are not much better than electric resistance heating in very cold climates. Choose an air-source heat pump with an HSPF above 8.0, or look into GeoExchange heat pump options

3. When buying a new furnace or boiler, look for a high AFUE. Gas furnaces are available with AFUEs as high as 96%. Gas boilers, oil furnaces, and oil boilers are available with efficiencies up to about 90%. In very cold climates, purchase a very high-efficiency unit, even though it may cost substantially more than an 80–85% efficient unit. In milder climates, the payback period on a very high-efficiency unit may be too long; a cheaper, 82% efficient unit may be a better buy. The 1992 government AFUE standard requires a minimum 78% rating for furnaces and 80% rating for boilers. Buy a sealed combustion model with outside air supplied directly to the burner and exhaust gases vented directly outside.

4. Older heating systems that are still in good shape can usually be upgraded. Most older burners are oversized, especially in homes that have been weatherized, and they don't burn fuel as efficiently as the new models. With oil systems, you can downsize by installing a smaller nozzle. Installing a new flame retention burner downsizes and otherwise improves the efficiency of the burner at the same time. With some gas systems, you can downsize by modifying the orifice and installing new baffles. You can downsize and improve the efficiency of a gas burner by installing a high-efficiency power burner.

5. Other modifications that can improve efficiency on older systems include barometric dampers in the flues (oil only), modulating aquastats or time delay relays (boilers only), automatic flue dampers, adjustable radiator vents and valves, and clock thermostats.

6. If your boiler provides domestic hot water with a tankless coil, its efficiency is quite low in the summer. The best solution is to install an indirect water heater, which operates off your boiler. You might also want to consider a solar water heater for summertime use, keeping the tankless coil for backup use when the heating system is on.

7. All heating systems must have regular maintenance. Residents should change filters on warm air furnaces, make sure radiators and registers are not blocked, check steam boiler water levels, and maintain radiators. A heating system technician should clean and tune oil burners each year and gas burners every two to three years. The technician should measure efficiency before and after the tune-up. Duct sealing can improve the efficiency of an older warm-air furnace or ensure that a new system operates optimally.

8. Keep your thermostat set at the lowest comfortable level. Don't heat unused rooms. Turn the thermostat down at night and when no one is home. Increase comfort levels at lower thermostat settings by maintaining humidity, stopping cold air leaks, and covering windows at night. Set aquastats to lower temperatures in milder weather. Set back the fan thermostat on warm air furnaces.

CHAPTER 5
Cooling Systems

$\mathbf{M}$ore than two-thirds of all U.S. households have air conditioners, whose energy consumption amounts to almost 5% of all the electricity produced in the U.S. for all purposes, at a cost to homeowners of over $10 billion. That amount of electricity results in the release of roughly 100 million tons of carbon dioxide (CO_2) per year, or an average for homes with air conditioning of one and a half per year. Furthermore, air conditioning use in the U.S. has more than doubled just since 1981. Seventy-five percent of new homes are being outfitted with central air conditioning systems, including over 50% of new homes in the Northeast.

A switch to high-efficiency air conditioners and implementation of measures to reduce cooling loads in homes can reduce this energy use by 20-50%.

After briefly describing how air conditioners work, this chapter covers selection of a new system, upgrading existing air conditioners, operating air conditioners for maximum efficiency, and, finally, strategies to reduce the need for air conditioning. The information on reducing cooling loads can be valuable to you whether or not you use an air conditioner. By following the strategies outlined, many homes will be able to eliminate the need for air conditioning on all but the hottest days and with no sacrifice in comfort.

AIR CONDITIONING BASICS

Air conditioning, or cooling, is more complicated than heating. Instead of using energy to create heat, air conditioners use energy to take heat away. The most common air conditioning system uses a compressor cycle (similar to the one used by your refrigerator) to transfer heat from your house to the outdoors.

Picture your house as a refrigerator. There is a compressor on the outside filled with a special fluid called a refrigerant. This fluid can change back and forth between liquid and gas. As it changes, it absorbs or releases heat, so it is used to "carry" heat from one place to another, such as from the inside of the refrigerator to the outside, or from inside the house to the outside. Simple, right?

Well, no. And the process gets quite a bit more complicated with all the controls and valves involved. But its effect is quite remarkable. An air conditioner takes heat from a cooler place and dumps that heat in a warmer place, seemingly working against the laws of physics. What drives the process, of course, is electricity—quite a lot of it, in fact.

TYPE OF AIR CONDITIONERS

There are three common types of air conditioners: room air conditioners, central air conditioners, and electric heat pumps. Room air conditioners are available for mounting in windows or direct mounting in walls, but in each case they work the same way, with the compressor located outside. Room air conditioners are sized to cool just that one room, so a number of them may be required for a whole house.

Central air conditioners are designed to cool the entire house. The large compressor unit is located outside, and the inside coils cool air that is distributed throughout the house via ducts. Often, the same duct system is used for forced warm air heating systems and central air conditioners.

Heat pumps are like central air conditioners, except that the cycle can be reversed and used for heating during the winter months. Using heat pumps for heating is described in Chapter 4.

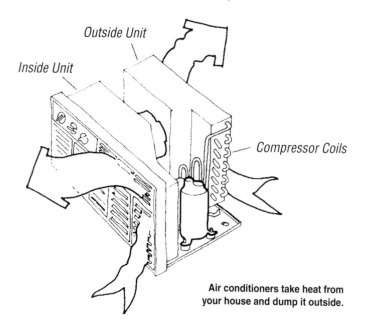

Outside Unit

Inside Unit

Compressor Coils

Air conditioners take heat from your house and dump it outside.

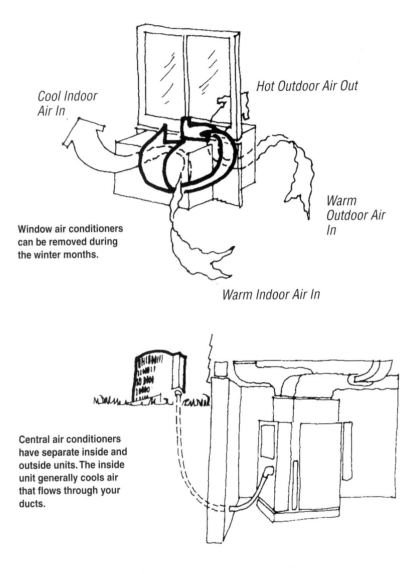

Cool Indoor Air In

Hot Outdoor Air Out

Window air conditioners can be removed during the winter months.

Warm Outdoor Air In

Warm Indoor Air In

Central air conditioners have separate inside and outside units. The inside unit generally cools air that flows through your ducts.

EVAPORATIVE COOLERS

In addition to these common types of air conditioners, there are several other types used very successfully in certain parts of the country. Evaporative coolers are practical in very dry areas, such as the Southwest. Sometimes called swamp coolers, they work by blowing house air over a damp pad or by spraying a fine mist of water into the house air. The dry air evaporates moisture and cools off. You've experienced this process—it's why a breeze makes you feel cold when you get out of a swimming pool. A direct evaporative cooler adds moisture to a house.

An indirect evaporative cooler is a little different in that the evaporation of water takes place on one side of a heat exchanger. House air is forced across the other side of the heat exchanger where it cools off but does not pick up moisture. Like the direct evaporative cooler, though, this system depends on very dry air to operate.

SELECTING A NEW AIR CONDITIONER

Choosing an air conditioner (or heat pump) is an important decision. Buying an inefficient model will lock you into high electric bills for years to come. The same issues apply whether you're buying your first air conditioner or replacing an existing one, though your choices may not be as numerous with replacement systems.

If you live in a hot, arid region, such as the Southwest, look into evaporative coolers. For the rest of the country, compressor-driven air conditioning systems are about the only choice, other than natural cooling (see discussion later in this chapter). This discussion will focus on conventional compressor-type air conditioners.

Choosing the Right Type of Air Conditioner

The type of air conditioner you need depends in large part on your climate and cooling loads. In small homes and those with modest cooling needs, room air conditioners often make the most sense. In fact, in a small, highly insulated house, even the smallest central air conditioner may be too large. If you are considering room air conditioners, you will need to decide between units that mount in the window and those that are built into the wall. Wall-mounted units are often a better choice, both for aesthetic and practical reasons, though they will cost more to install, because an opening has to be cut through the wall. Window air conditioners are harder to seal, they block views and light, and they prevent the use of the window for natural or forced ventilation.

Central air conditioners have a number of advantages. They are out of the way, quiet, and convenient. If you already have a forced-air heating system, you may be able to tie into the existing duct work. Whether or not your existing ducting will work for air conditioning depends on its size and your relative heating and cooling loads. Ask your air conditioning service technician. Plus, central air conditioners are more efficient (see **Efficiency** below).

Heat pumps, though more expensive, provide heat in addition to air conditioning all in one unit. If you already have a satisfactory gas or oil heating system and have decided to add air conditioning, it usually doesn't make sense to consider a heat pump, because even a high-efficiency heat pump will probably be more expensive to operate than your gas or oil heating system (see the discussion on heat pumps in Chapter 4). If you currently have electric resistance heat, however, and you live in a relatively warm climate (winter temperatures seldom dropping below 30°F), a heat pump may be a good choice.

If you're unsure about which type of air conditioner makes the most sense for your house, ask for opinions and bids from several local air conditioning installers.

Sizing the System

No matter what type of system you choose, make sure that it is sized properly. Most air conditioners are rated in Btu/hour, but central air conditioners and heat pumps may also list cooling capacity by the ton. One ton is equivalent to 12,000 Btu/hour. With air conditioning systems, equipment cost is much more proportional to size than it is with heating equipment. Doubling the cooling output nearly doubles the cost, so it makes a lot of sense to be very careful with sizing. Don't let a salesman convince you to buy an oversized system. In addition to the higher cost for an oversized system, it will run only for short periods, cycling on and off, which will increase electricity use and decrease the unit's overall efficiency. If it just runs for short periods of time, it also won't do as good a job dehumidifying the air (see **Dehumidification** below).

Find a qualified air conditioning technician or energy auditor to determine your cooling load. Do not rely on simple rules-of-thumb by air conditioner salesmen, but insist on thorough analysis, including local climate information and calculations of heat gain through windows and walls.

You can save a lot of money by reducing your cooling loads as much as possible before buying the system. (See **Reducing the Need for Air Conditioning** later in this chapter.) Make sure conservation efforts are taken into account when the technician is figuring out how large a system you need.

With heat pumps, proper sizing can be especially difficult, because the same unit is used for both cooling and heating. A heat pump sized for heating loads in a cold climate will be considerably oversized when it comes to cooling, and a heat pump that is sized for cooling loads in a warm climate will tend to be oversized when it comes to heating. If the heating load is larger than the cooling load, some heat pump salespeople will recommend sizing the heat pump for cooling and then adding enough electric resistance heat to make up the difference in the winter. In such a situation, it generally makes more sense to size the heat pump to provide all of the heating requirements in average winter conditions, even though it will mean a larger and somewhat more expensive model. A good heat pump technician should be able to help you choose the best compromise between cooling and heating capacity.

Efficiency

Efficiency is just as important with air conditioning systems as it is with heating systems. Central air conditioners and heat pumps operating in the cooling mode are rated according to their seasonal energy efficiency ratio (SEER), which is the seasonal cooling output in Btu divided by the seasonal energy input in watt-hours for an average U.S. climate. Many older central air conditioners have SEER ratings of only 6 or 7. The average central air conditioner sold in 1988 had a SEER of about 9. The national efficiency standard for central air conditioners now requires a minimum SEER of 10, and to qualify for an Energy Star® label requires a SEER of 12 or higher.

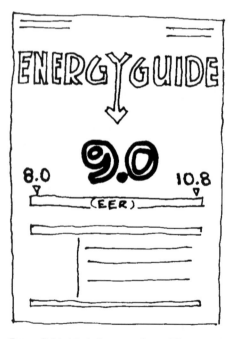

The efficiency of room air conditioners is measured by the energy efficiency ratio (EER), which is the ratio of the cooling output (in Btu) divided by the power consumption (in watt-hours). A typical new room air conditioner has an EER of about 9. The first national appliance efficiency standards for room air conditioners took effect in 1990. The minimum efficiency varies depending on the design and cooling capacity of each unit. On average, the 1990 standard requires a minimum EER of about 8.6. New standards will go into effect in October 2000 requiring an average EER of about 10.

EnergyGuide labels for room air conditioners list the Energy Efficiency Ratio or EER.

When you're shopping for air conditioners, look for SEER ratings over 12 for central air conditioners and EER ratings over 10 for room air conditioners. High-efficiency units generally cost more, but in hot climates more efficient units pay for themselves over a few years through reduced electricity bills. Central air conditioners are usually more efficient than room air conditioners, and in general, larger capacity air conditioners have higher efficiency. However, don't buy a larger system than you need just because it has higher efficiency (see discussion on sizing above).

Other energy-saving features to look for include a fan-only switch, which will enable you to use the unit for nighttime ventilation and substantially reduce air conditioning costs (see discussion on ventilation below); a filter check light to remind you to check the filter after a pre-determined number of operating hours; an automatic delay fan switch to turn off the fan a few minutes after the compressor turns off; and quiet operation (sound ratings are generally not listed by manufacturers, but with room air conditioners, ask if you can listen to the different models in operation).

Dehumidification

Air conditioners remove a certain amount of moisture from the air because the room air is forced past cold coils. Water vapor from the air condenses out on the coils the same way moisture from the air condenses on a glass of ice water on a hot, humid day. This water drains outside through a condensate drain.

Lowering the humidity in this way is both good and bad. You feel more comfortable at lower humidity levels, so the dehumidification contributes to cooling. But when water vapor condenses into liquid, it releases stored heat, reducing the efficiency of the air conditioner. One of the ways manufacturers have boosted the efficiencies of air conditioners in recent years is by keeping the condenser coils somewhat warmer, thus reducing condensation. Some of the new high-efficiency air conditioners, therefore, do not dehumidify air as effectively. This can be a problem, especially in the humid Southeast.

High-efficiency air conditioning systems can get around the dehumidification problem by including variable-speed or multispeed blowers. High-speed operation leads to high efficiencies but low dehumidification. Lower speeds reduce efficiency somewhat, but increase dehumidification. Lower speeds are used during very humid weather; the rest of the time a more efficient higher speed is used.

If you live in a humid climate, look for air conditioner models that are effective at removing moisture. Although there is no industry standard for rating the effectiveness at removing moisture, most literature does list water removal in pints per hour, which will help you compare one model to another. Choose a model with a variable-speed fan to aid in dehumidification if you're buying a very high-efficiency air conditioner, even though you might have to pay a little more for it. Several manufacturers have models with a variable-speed blower controlled by a humidistat, automatically reducing fan speed at high humidities. Also try to keep moisture out of the house (see **Operation and Maintenance** below).

THE MOST ENERGY-EFFICIENT AIR CONDITIONERS

The charts below list the most energy-efficient air conditioners on the market today. These are divided into room air conditioners and central air conditioners. (For heat pump efficiencies, see Chapter 4. The heat pump listings in that chapter include both heating efficiency and cooling efficiency.)

The most efficient room air conditioners have EERs ranging from 10.0 to 11.7. Most of these models are available for either window mounting or through-the-wall mounting. All of the models listed below meet the new minimum standards for room air conditioners, which will take effect in October, 2000.

MOST EFFICIENT ROOM AIR CONDITIONERS				
Brand	Model	Volts	Capacity	EER
Less than 6,000 Btu/hr cooling capacity				
Friedrich	SQ05J10A	115	5,400	11.0
Amana	5M11TA	115	5,000	10.0
Carrier	UCA051B	115	5,200	10.0
Whirlpool	ACQ052XH0	115	5,200	10.0
Frigidaire	FAC056F7A	115	5,450	10.0
Gibson	GAC056Y7A	115	5,450	10.0
Kenmore	253.975055	115	5,450	10.0
White-Westinghouse	WAC056F7A	115	5,450	10.0
Frigidaire	FAC056G7A	115	5,500	10.0
Frigidaire	FAC056H7A	115	5,500	10.0
Gibson	GAC056G7A	115	5,500	10.0
Kenmore	253.7005599	115	5,500	10.0
Kenmore	253.7805589	115	5,500	10.0
White-Westinghouse	WAC056G7A	115	5,500	10.0
White-Westinghouse	WAC056H7A	115	5,500	10.0
General Electric	AMH06LA	115	5,800	10.0
Panasonic	CW-606TU	115	5,800	10.0
Quasar	HQ2062KH	115	5,800	10.0
Signature 2000	KMJ-5817	115	5,800	10.0

MOST EFFICIENT ROOM AIR CONDITIONERS (cont.)

Brand	Model	Volts	Capacity	EER
6,000 to 7,999 Btu/hr cooling capacity				
Cold Point	EP08RSA	115	7,440	10.9
Friedrich	SQ06J10A	115	6,300	10.8
Friedrich	SQ07J10A	115	7,100	10.3
Frigidaire	FAB067W7B	115	6,100	10.0
Gibson	GAB067F7B	115	6,100	10.0
Kenmore	253.975061	115	6,100	10.0
White-Westinghouse	WAB067F7B	115	6,100	10.0
Friedrich	YQ06J10A	115	6,200	10.0
Sharp	AF-T706X	115	6,500	10.0
Amana	7M11TA	115	6,600	10.0
Kenmore	596.78079890	115	6,600	10.0
Danby	DAC7059	115	7,000	10.0
Daewoo	DWC-070C	115	7,390	10.0
General Electric	A*H08FA	115	7,800	10.0
Panasonic	CW-806TU	115	7,800	10.0
Quasar	HQ2082KH	115	7,800	10.0
Signature 2000	KMJ-5819	115	7,800	10.0
8,000 to 9,999 Btu/hr cooling capacity				
Friedrich	YS09J10	115	9,000	11.5
Friedrich	SS09J10A	115	9,200	11.5
Friedrich	SS08J10A	115	8,200	10.8
White-Westinghouse	WAK083F7V	115	8,000	10.5
Friedrich	SQ08J10B	115	8,000	10.0
General Electric	AGN08FA	115	8,000	10.0
Goldstar	LW-B0813C*	115	8,000	10.0
Sharp	AF-*908X	115	8,500	10.0
Sharp	AF-T906XB	115	8,500	10.0
Carrier	TCA081P	115	8,600	10.0
Amana	9M32P*E	230	9,100	10.0
Amana	9M32P*EH	230	9,500	10.0
Amana	B9M32PAEH	230	9,500	10.0
10,000 to 12,999 Btu/hr cooling capacity				
Friedrich	SS10J10A	115	10,200	11.7
Friedrich	RS10J10	115	10,000	11.0
Friedrich	RS12J10	115	12,000	10.5
Friedrich	SS12J10A	115	12,000	10.5
Friedrich	ES12J33	230	12,000	10.5
Friedrich	SS12J30A	230	12,000	10.5
Friedrich	KS10J10	115	10,000	10.3
Bryant	462AJH012BA	230	12,000	10.2
Carrier	XHA123D	230	12,000	10.2
13,000 to 16,999 Btu/hr cooling capacity				
Mitsubishi Electronics	MSH15NN	208	14,600	10.6
Friedrich	SM14J10B	115	14,000	10.5
Friedrich	RM15J10	115	14,500	10.5

MOST EFFICIENT ROOM AIR CONDITIONERS (cont.)				
Brand	Model	Volts	Capacity	EER
13,000 to 16,999 Btu/hr cooling capacity (cont.)				
Frigidaire	FA*15**1A	115	15,000	10.4
Gibson	GA*15**1A	115	15,000	10.4
Kenmore	253.'7815689	115	15,000	10.4
Kenmore	253.7915699	115	15,000	10.4
Kenmore	253.975156	115	15,000	10.4
White-Westinghouse	WA*15**1A	115	15,000	10.4
Mitsubishi Electronics	MSH17NW	208	16,200	10.3
Comfort Aire	R-141F-*	115	14,000	10.2
Crosley	CA14WC9*	115	14,000	10.2
Whirlpool	ACQ142XH*	115	14,000	10.2
Bryant	462AJC015BA	230	15,000	10.2
Carrier	X*B153D	230	15,000	10.2
Carrier	XCA141D	115	13,800	10.1
Friedrich	KM14J10	115	14,000	10.1
17,000 to 19,999 Btu/hr cooling capacity				
Amana	18M23TA	230	18,000	10.2
Kenmore	596.7818989	230	18,000	10.2
Friedrich	EM18J34A	230	18,500	10.0
Friedrich	RM18J30	230	18,500	10.0
Friedrich	SM18J30A	230	18,500	10.0
Kenmore	596.78189890	230	18,000	10.0

Manufacturers now offer a large number of very efficient central air conditioning systems. The models listed on the next pages are representative of manufacturers' top model lines. Bear in mind that a SEER of 12 or higher identifies a very efficient system, and a SEER of 12 or better earns the ENERGY STAR® distinction.

MOST EFFICIENT CENTRAL AIR CONDITIONERS				
Brand	Condensing Unit	Blower Coil	Btu/hr	SEER
Cooling Capacity: Approximately 2 tons				
Amana	RCE24A2A	CHA24T*C+GUIV070DX40	25,400	16.00
Bryant	556AN024-A/C	FV4NF003	25,000	15.60
Fraser-Johnston/ Luxaire	HABE-F024	GBFD046S17+NAVSB12+ 1TV0701	26,200	15.50
Trane	TTY024B	TWE040E13	25,800	15.50
York	H1RE024S06	G2FD046S17+N1VSB12+1 TV0701	26,200	15.50
Airquest/Heil/Tempstar	CA9624*KB*	EX*42J****+EV16J22****	25,000	15.00

There are many models with SEERs above 12. All above 12 are at least 20% more efficient than minimum standards.

MOST EFFICIENT CENTRAL AIR CONDITIONERS (cont.)

Brand	Condensing Unit	Blower Coil	Btu/hr	SEER
Cooling Capacity: Approximately 2 tons (cont.)				
American Standard	7A4024A	TWE040E13	26,400	15.00
Int'l Comfort Products	AJ024GA*	EX*42J****+EV16J22****	25,000	15.00
Carrier	38TSA02430/32	CJ5A/CK5AW024=R410A TXV+58MVP080-14	24,000	15.00
Goodman/Janitrol/ Kenmore	CKQ24-1	U47/UC47+GMPE075-3	23,000	15.00
Goodman/Janitrol/ Kenmore	CKQ24-1	AE24-XX	23,000	15.00
Lennox	HS27-024-1P	CB31MV-41-1P	26,000	15.00
Trane	TTX024C	TUD100R9V5+TXC037S3	25,400	14.60
York	H2DH024S06	G2FD036S17+N1VSB12	24,400	14.10
Cooling Capacity: Approximately 2.5 tons				
Amana	RCE30A2B	CCH30TCD+GUIV115DX50	29,400	16.00
Bryant	556AN030-A/C	FV4ANF005	30,000	15.75
Bryant	556AN030-A/C	FK4CNF005+R410A TXV	30,000	15.75
Carrier	38TSA03030/32	FV4ANF005	30,000	15.75
Carrier	38TSA03030/32	40FKA/FK4CNF005+R410A TXV	30,000	15.75
Lennox	HS27-030-1P	CB31MV-41-1P	29,200	15.50
Fraser-Johnston/ Luxaire	HABE-F030	GBFD046S17+NAVSB12+ 1TV0701	30,000	15.35
Trane	TTY030B	TWE040E13	33,000	15.35
York	H1RE030S06	G2FD046S17+N1VSB12+ 1TV0701	30,000	15.35
Airquest/Heil Kenmore/Tempstar	CA9630*KC*	EX*42J****+EV16J22****	31,000	15.05
Int'l Comfort Products	AJ030GB*	EX*42J****+EV16J22	31,000	15.05
Cooling Capacity: Approximately 3 tons				
Trane	TTZ036A	TWE040E13	38,000	18.00
American Standard	7A6036A	TWE040E13	38,500	17.20
Amana	RCE36A2A	CCA48FDC+BBC36A2A+TXV	37,600	16.50
Carrier	38TDA03630	40FKA/FK4CNF003	37,000	16.00
Carrier	38TDA03630	CJ5A/CK5A/CK5B*0**+TXV+ 58U(H,X)V1*0-20	37,000	16.00
Bryant	556AN036-A/C	FV4ANB006	36,200	15.75
Carrier	38TSA03630/32	FV4ANB006	36,200	15.75
Lennox	HS21-411-1P/2P	CB31MV-51-1P	38,000	15.75
Lennox	HS27-036-1P	CB31MV-51-1P	35,600	15.60
Trane	TTY036B	TDD100R9V5+TXC037S3	37,200	15.05
Cooling Capacity: Approximately 3.5 tons				
Carrier	38TDA03630	40FKA/FK4CNB006	40,000	17.10
Amana	RCE42A2A	CHA54TCC+BBC60A2A	42,000	16.00
Bryant	598AN036-C	FK4CNB006	40,000	16.00

There are many models with SEERs above 12. All above 12 are at least 20% more efficient than minimum standards.

MOST EFFICIENT CENTRAL AIR CONDITIONERS (cont.)				
Brand	Condensing Unit	Blower Coil	Btu/hr	SEER
Cooling Capacity: Approximately 3.5 tons (cont.)				
Coleman Evcon/				
Guardian	FRCS0421CE	G*FD060S24+N*VSD20+TXV	41,500	15.50
Fraser-Johnston/				
Luxaire	HBBE-F042	GBFD060S24+N1VSD20	41,500	15.50
Trane	TTY036B	TWE040E13	39,000	15.50
York	H2RE042S06	G2FD060S24+N1VSD103	41,500	15.50
Lennox	HS27-042-1P	CB31MV-65-1P	43,000	15.35
American Standard	7A4036A	TWE040E13	40,000	15.30
Bryant	556AN042-A/C	FV4ANB006	42,000	15.10
Carrier	38TSA04230/32	FV4ANB006	42,000	15.10
Cooling Capacity: Approximately 4 tons				
Trane	TTZ048A	TWE065E13	49,500	16.05
American Standard	7A6048A	TWE065E13	49,000	16.00
Amana	RCE48A2A	CHA60TCC+BBC60A2A	49,000	15.50
Coleman Evcon/				
Guardian	FRCS0481CE	G*FD060S24+N*VSD20+TXV	47,500	15.50
Coleman Evcon/				
Guardian	FRCS0481CE	F*FV060N+TXV	47,500	15.50
Fraser-Johnston/				
Luxaire	HBBE-F048	GBFD060S24+N1VSD20+ 1TV0703	47,500	15.50
Fraser-Johnston/				
Luxaire	HBBE-F048	FBFV060+1TV0703	47,500	15.50
York	H2RE048S06	G2FD060S24+N1VSD20+ 1TV0703	47,500	15.50
Carrier	38TDA04830	40FKA/FK4CNB006	51,000	15.30

There are many models with SEERs above 12. All above 12 are at least 20% more efficient than minimum standards.

BUYING AND INSTALLING AN AIR CONDITIONING SYSTEM

Room air conditioners can be purchased for as little as a few hundred dollars, while large central air conditioners and heat pumps can cost as much as $5,000. If any modifications need to be made in the ducting system for a central air conditioner, that can add substantially to the cost. It pays to shop around and get bids from a number of different contractors.

When evaluating bids, be sure to consider what you are getting for the price. Does the system come with a warranty? If so, how long is it? Air conditioner warranties range from one year for complete parts and labor to five years for the compressor. Some manufacturers are now offering ten-year warranties on the compressors. If an existing system

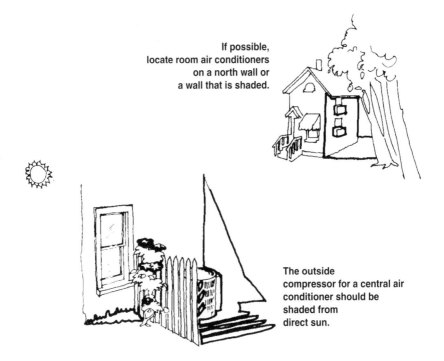

If possible, locate room air conditioners on a north wall or a wall that is shaded.

The outside compressor for a central air conditioner should be shaded from direct sun.

is being replaced, will the old unit be hauled away? Does the air conditioning contractor offer a service plan, and is it affordable? Make sure the contractor has been in business for a while and is fully bonded and insured. If you are not familiar with the company, ask for some local references and follow them up.

To maximize efficiency, the outside part of a central air conditioner, the compressor, should be located in a cool, shaded place. The best place is usually on the north side of the house under a canopy of trees or tall shrubs. However, it shouldn't be choked by vegetation; the compressor needs unimpeded air flow around it to dump waste heat effectively. Never place the compressor on the roof or on the east or west side unless it is completely shielded from the summer sun, because sunlight shining on it will heat it up and reduce its efficiency at dumping heat.

Also, the compressor may be somewhat noisy. Try to keep it some distance from a patio or bedroom window. If you're concerned about noise, ask to see (and hear) one in operation before buying it.

With heat pumps, location of the outside unit is more complicated. Because a heat pump is used for both cooling and heating, it usually makes more sense to locate the compressor on the south side—especially in colder climates, and shade it with a sunscreen or tall annuals during the summer. In the winter, when the compressor is trying to extract heat, a southern location will allow it to absorb solar energy, which will boost its heating efficiency.

The location of room air conditioners is constrained by available walls and windows. To perform most efficiently, these units should be out of direct sunlight; if you have a choice of walls, the north is best and the south is second best; avoid east or west walls if at all possible.

UPGRADING EXISTING AIR CONDITIONERS

The compressor units of most air conditioners have an average lifetime of only around 10 to 12 years. By carefully following proper maintenance procedure, a quality model may hold up twenty years, but don't expect the kind of lifetime you get with boilers and furnaces.

If you have an old air conditioner—more than ten years old—chances are pretty good that it's also inefficient. A ten-year-old central air conditioner probably has a SEER rating between 7 and 8, compared with the best new models that are up to twice as efficient. It will definitely pay to replace it, but you may not have to replace the entire air conditioner. Sometimes just the outdoor compressor component needs to be replaced, though it may be hard to find high-efficiency parts for low-efficiency models.

If you're replacing just the compressor, though, make sure that the new outdoor unit is compatible with the indoor blower coil. The highly efficient outdoor unit will not reach its rated efficiency if it is not properly matched to the indoor unit. An air conditioning service technician should be able to help you match units effectively.

OPERATION AND MAINTENANCE

Using natural or forced ventilation at night, while keeping the house closed up tight during hot days, is less expensive than operating your air conditioner (see **Reducing the Need for Air Conditioning** below). Use air conditioning only when ventilation is inadequate. Don't cool unoccupied rooms, but don't shut off too many registers with a central system either, or the increased system pressure may harm the compressor. If your air conditioner has an outside air option, use it sparingly. It is far more economical to recirculate and cool the indoor air than to cool the hot outdoor air down to comfortable temperatures. Always keep all doors

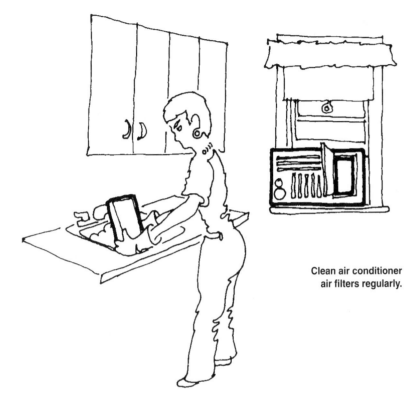

Clean air conditioner
air filters regularly.

and windows closed when operating an air conditioner. Do not operate a whole-house fan or window fans while using the air conditioner.

You will probably be comfortable with the thermostat set at about 78°F, but ceiling fans can increase your comfort range. You will save 3-5% on air conditioning costs for each degree that you raise the thermostat. You can also increase comfort at warmer temperatures by reducing humidity; use a bathroom exhaust fan when you shower, don't dry firewood in your basement, and don't vent your clothes dryer inside.

Air conditioners and heat pumps need regular maintenance in order to perform at peak efficiency. Clean the air filters on room air conditioners monthly. They should never be allowed to get dirty enough to impede air flow, as this could cause damage to the unit. The condenser should be cleaned by a professional every other year, or even yearly in dusty locations.

Central air conditioning units should be inspected, cleaned, and tuned by a professional once every two to three years. This will extend the life of the unit and reduce electricity consumption. Check with your service technician about the proper maintenance schedule for your unit.

During service of your unit, its refrigerant may need recharging. It needs to be charged correctly. A 20% undercharged system can operate at 20% lower efficiency. However, an overcharged system not only reduces operating efficiency, it can cause damage to the unit and reduce the lifetime of the system. Also, because refrigerants damage the ozone layer, it is important that the refrigerant not be leaked to the environment; it can and should be recycled.

The service technician should also measure airflow over the indoor coil. Recent studies show that inadequate airflow is a common problem and average air flow rates tend to degrade over time due to poor maintenance. Correction of airflow rates can improve efficiency by 5-10%.

Even if an air conditioner or heat pump is installed and maintained with adequate air flow and the appropriate level of refrigerant, the unit will not operate efficiently if the duct system is in poor condition. Duct sealing can reduce cooling energy use by 10-15%. (See **Modifications by Heating Service Technicians** in Chapter 4 for more on duct sealing).

The power to a central unit should be shut off when the cooling season ends, otherwise the heating elements in the unit could consume energy all winter long. Flip the circuit breaker to turn it off if the unit doesn't have a separate switch. Turn the power back on at least one day before starting up the unit in order to prevent damage to the compressor.

REDUCING THE NEED FOR AIR CONDITIONING

Even with the most efficient air conditioners, it makes a great deal of sense to do everything you can to reduce air conditioning loads. The following conservation measures are often so effective that houses in the northern third of the country and in mountainous regions can get by without air conditioning on all but the very hottest days. If you're planning to buy a new system, reducing the cooling load will save you a lot of money right away be letting you buy a smaller, less expensive system.

Human Comfort

In looking at how air conditioning costs can be reduced, it helps to understand human comfort. The standard human comfort range for light clothing in the summer is between 72°F and 78°F and between

35% and 60% relative humidity, according to the American Society for Heating, Refrigerating and Air-Conditioning Engineers (ASHRAE). The comfort range can be extended to 82°F with modest air movement, as might be provided by ceiling fans, for example. Often the house can be kept within this range using little or no mechanical air conditioning.

To extend the comfort range to 82°F, you need a breeze of about 2.5 ft/sec or 1.7 mph. A slow-turning ceiling-mounted paddle fan can easily provide this air flow. The following table gives the necessary fan blade diameters for various size rooms. Fans should have multiple speed settings so that air flow can be reduced at lower temperatures.

TABLE 5.1

CEILING FAN SIZING

Room area (sq ft)	Minimum fan diameter (inches)
100	36
150	42
225	48
375	52
400+	2 fans

Source: Don Abrams, *Low Energy Cooling.*

Sources of Unwanted Heat

There are three major sources of unwanted heat in your house during the summer: heat that conducts through your walls and ceiling from the outside air, waste heat that is given off inside your house by lights and appliances, and sunlight that shines through your windows. These are described below, along with techniques to reduce them.

1. *Heat gain through your walls and ceiling.* Whenever the outdoor temperature is higher than the indoor temperature, heat will conduct through the walls and ceiling of the house to the interior. Warm air will blow into the house through cracks.

To reduce these gains, you can insulate and tighten your house. One of the most cost-effective energy conservation measures, for both heating and cooling, is to add extra ceiling insulation. Increase its depth to a full 12". If you don't have wall insulation, have cellulose or

Common heat sources in a house.

fiberglass blown into the walls by a qualified insulation contractor. Tighten up your house to reduce infiltration. You might also want to install a radiant barrier in the attic to cut down on summer heat gain. If properly installed, a radiant barrier can reduce cooling costs to some extent. New light-colored roofing materials and roof coatings can also reduce cooling loads. An energy auditor can help you decide which measures make the most sense for your house and how much they will cost (see Chapter 2).

2. *Waste heat from appliances and lights.* Most of the energy used for lights, refrigerators, stoves, washers and dryers, dishwashers, and other household appliances eventually ends up as waste heat that will raise the interior temperature of your house.

The best solution is to buy energy-efficient products. Energy-efficient appliances and lights produce far less waste heat. Standard incandescent

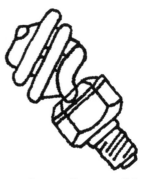

Compact fluorescent lights add less heat to the house than incandescent lights.

light bulbs, for example, emit 90% of their energy as heat—only 10% as light. Compact fluorescent lights produce only a fraction of the heat (see Chapter 11). In some cases, you can delay heat-producing tasks, such as dishwashing, until the cooler evening hours. You might also consider relocating a freezer to the basement or garage, where it won't contribute its waste heat to your living space. And by planning your meals carefully, you can minimize use of the oven on the hottest days.

3. *Solar gain through windows.* Sunlight shining in windows, particularly those on the east and west sides of the house, usually adds the largest amount of unwanted summertime heat. In addition, the sun heats up the roof and walls of the house, increasing heat conduction to the interior. With no shading of east and west windows, the interior temperature of a typical house could rise as much as 20°F on a hot day, either making your air conditioner work a lot harder, or making you a lot less comfortable.

The best way to eliminate solar gain is to provide effective shading. Use horizontal trellises above east- and west-facing windows, which

Attic ventilation is a very important way to get rid of unwanted summer heat.

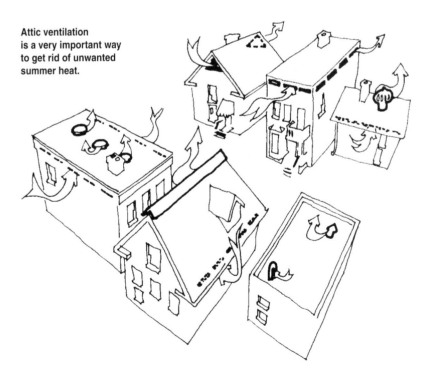

Trees are a very
effective way to shade
a house on the
east and west.

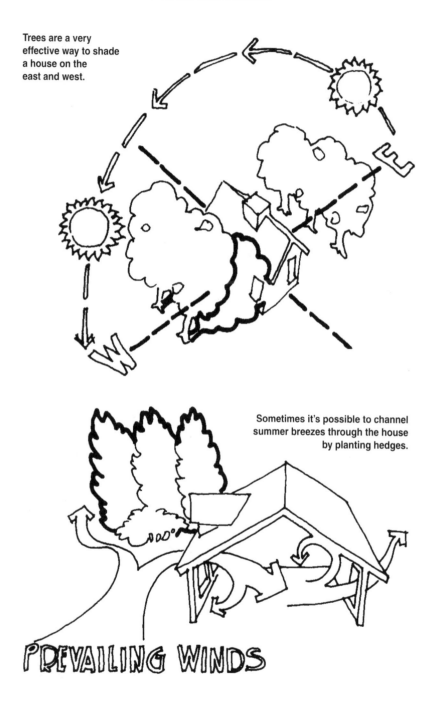

Sometimes it's possible to channel
summer breezes through the house
by planting hedges.

PREVAILING WINDS

collect more summer sun than the others. Plant tall trees (prune lower branches so as not impede summer breezes), or use awnings wider than the windows to provide shade. If you have a choice, place porches, sheds, and garages on east and west walls to provide further shading. Unless you have extensive areas of south glass, the south wall should not require summer shading because the summer sun is at too high an angle to cause much of a problem. Planting trees in front of south windows will block beneficial winter gains.

If you're replacing windows, put in high-performance windows with low-e glazings that look perfectly clear yet block out a large percentage of unwanted heat gain (see Chapter 3). It makes sense to install windows with low solar heat gain coefficients (SHGC) on west and east walls, where heat gain from the sun is the greatest in the summer. Windows with high SHGC make more sense on the south, especially when you want to benefit from passive solar heating during the winter months. Another way to reduce solar gain through windows is to install drapes with light-colored linings or operable blinds that will reflect sunlight back outside. Vertical blinds are particularly effective on east- and west-facing windows. Also choose lighter colors for roofs and walls to reflect sunlight and reduce conductive heat gain.

Getting Rid of Unwanted Heat through Ventilation

Natural or mechanical ventilation can help reduce air conditioning costs in every area of the country. In the northern United States and in mountainous areas, ventilation can often totally eliminate the need for air conditioning. In the rest of the country, ventilation can provide comfort over much of the cooling season, especially during the spring and fall months.

In order for ventilation to be most effective, the temperature of the incoming air should be 77°F or lower, making this strategy most effective at night and on cooler days. Keep the house closed up tight on hot days and try to limit unwanted heat gains as outlined above, then ventilate the house at night. In breezy locations, natural ventilation can be provided simply by opening screened windows. Plantings and fences can be used to help funnel breezes towards your house. If there isn't much wind, you'll need to provide mechanical ventilation with either window fans or a whole-house fan.

Window Fans. Window fans for ventilation are a reasonable option if they are used properly. They should be located on the leeward (downwind) side of the house facing out. A window should be open in each room. Interior doors must remain open to allow air flow. Window fans will not work as well in houses with long, narrow hallways or those with small rooms and many interior partitions. Window fans can be noisy, especially on high settings, but they are inexpensive.

Whole-House Fans. A whole-house fan is a more convenient option than window fans and may cost no more than three or four window fans. Mounted in a hallway ceiling on the top floor, the fan sucks air from the house and blows it into the attic. The fan is usually covered on the bottom by a louvered vent. (To reduce heat loss through the fan during the wintertime, the en-tire assembly should be installed within an insulated, weather-stripped box with a removable or hinged lid.)

Window fan.

The fan should have at least two speeds, with the highest one capable of changing the entire volume of air in the house very quickly. Because the fan blows air into the attic, the attic must have sufficient outlet

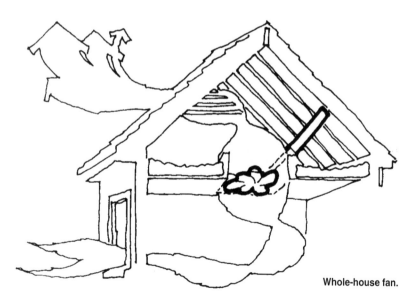

Whole-house fan.

vents. The free vent area, including soffit vents, ridge vents, and gable-end vents, should be twice the free vent area of the fan opening. (Free vent area is a measure of the area of the vent opening minus the area blocked by screening and louvers.)

Before turning on the fan, be sure to open several windows in various areas of the house. If just one or two windows are open, the air flow through them will be intolerably high. For safety reasons, the fan should have manual controls (either in addition to or instead of automatic controls). A *fusible link*, which automatically shuts the fan down in case of fire, should be included for safety (the controls should be installed by a licensed electrician). The fan should be installed carefully; a loose installation can cause vibrations and excessive noise. Units with low-speed motors (700 rpm or less) will usually be less noisy. Ask for a demonstration.

The whole-house fan can be turned on as soon as the outdoor temperature drops about three degrees below the indoor temperature. The fan speed should be adjusted according to how quickly you want to cool the house down. Mechanical ventilation uses far less electricity than mechanical air conditioning.

RECOMMENDATIONS

1. Reduce the cooling load by employing cost-effective conservation measures. Provide effective shade for east and west windows. When possible, delay heat-generating activities such as dishwashing until evening on hot days.

2. Over most of the cooling season, keep the house closed tight during the day. Don't let in unwanted heat and humidity. Ventilate at night either naturally or with fans.

3. Use ceiling fans to increase comfort levels at higher thermostat settings.

4. In hot climates, plant shade trees around the house. Don't plant trees on the south if you want to benefit from passive solar heating in the winter.

5. If you have an older central air conditioner, consider replacing the outdoor compressor with a modern, high-efficiency unit. Make sure that it is properly matched to the indoor unit.

6. If buying a new air conditioner, be sure that it is properly sized. Get assistance from an energy auditor or air conditioning contractor.

7. Buy a high-efficiency air conditioner: for room air conditioners, the

EER should be above 10; for central air conditioners, look for a SEER above 12.

8. In hot, humid climates, make sure that the air conditioner you buy will adequately get rid of high humidity. Models with variable or multi-speed blowers are generally best. Try to keep moisture sources out of the house.

9. Try not to use a dehumidifier at the same time your air conditioner is operating. The dehumidifier will increase the cooling load and force the air conditioner to work harder.

10. Seal all air conditioner ducts with mastic, and insulate ducts that run through unheated basements, crawl spaces, and attics.

11. Keep the thermostat set at 78°F—or higher if using ceiling fans. Don't air-condition unused rooms.

12. Maintain your air conditioners properly to maximize efficiency.

CHAPTER 6
Water Heating

Next to heating or cooling, water heating is typically the largest energy user in the home. As homes have become more and more energy efficient during the past 20 years, the percentage of energy used for water heating has steadily increased. This chapter takes a look at the high-efficiency water heaters available and how you can reduce water heating costs with your present water heater.

TYPES OF WATER HEATERS

Storage Water Heaters

Storage water heaters are by far the most common type of water heater used today in this country. Ranging in size from 20 to 80 gallons and fueled by electricity, natural gas, propane, or oil, storage water heaters work by heating up water in an insulated tank. When you turn on the hot water tap, hot water is pulled out of the top of the water heater and cold water flows into the bottom to replace it. The hot water is always there, ready for use. Because heat is lost through the walls of the storage tank (this is called standby heat loss), energy is consumed even when no hot water is being used. New energy-ef-

Storage water heater.

ficient storage water heaters contain higher levels of insulation around the tank, substantially reducing standby heat loss.

Demand Water Heaters

Demand or instantaneous water heaters do not contain a storage tank. A gas burner or electric element heats water only when there is a demand for hot water. Hot water never runs out, but the flow rate (gallons of hot water per minute [gpm]) may be limited. By eliminating standby losses from the tank, energy consumption can be reduced by 10–15%. Before rushing out to buy a demand water heater, though, be aware that they aren't appropriate for every situation, and they may not even save that much energy and money.

The largest readily available gas-fired demand water heaters can supply only three gallons of hot water per minute with a temperature rise of 90°F (50° to 140°F, for example). However, such a sharp temperature rise may only be needed in commercial applications, where the hot water is mixed with cold. For home use, a rise of 60°–70°F is sufficient, bringing water from 50°F to 110°–120°F. Gas demand water heaters can

Demand water heater.

provide four to five gpm at this temperature. If you've installed a low-flow showerhead (see **Conserve Water** below) and won't need to be doing a load of laundry or dishes while someone is taking a shower, then 4–5 gpm might be fine. But if you have a couple of teenagers in the house, or if you need hot water for several tasks at the same time, a demand water heater might not be adequate.

Electric demand water heaters provide even less hot water. The largest models require 18 kilowatts at 240 volts yet provide less than 2 gpm at a 70°F temperature rise. Then again, an electric demand unit might make good sense in an addition or remote area of the house, thereby eliminating the heat losses through the hot water pipes to that area. Small electric demand water heaters can be placed under sinks to boost the temperature of incoming warm water. If you want to consider an electric unit, make sure your electrical wiring can handle the job before you make a purchase.

Demand water heaters make the most sense in homes with one or two occupants, and in households with small and easily coordinated hot water requirements. If you decide on a demand water heater, look for one that provides constant-temperature water at different flow rates—a feature called *modulating temperature control*. Otherwise, you and your family may find yourselves unhappy with fluctuating water temperatures—particularly if you have your own water system with varying pressure.

With gas-fired demand water heaters, keep in mind that the pilot light can waste a lot of energy. In gas storage water heaters, energy from the pilot light is not all wasted because it heats the water in the tank. This is not the case with demand water heaters. A 500 Btu/hour pilot light can consume 20 therms of gas per year, offsetting some of the savings you achieve by eliminating standby losses of a storage water heater. The majority of currently available demand water heaters do not have pilot-less ignition; one exception is a new AquaStar model which comes without a continuously burning pilot. To solve this problem, you can keep the pilot light off most of the time, and turn it on when you need hot water—a routine that should work fine in a vacation home, but not in a regular household.

Heat Pump Water Heaters

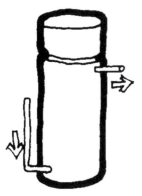

Heat pump water heaters are more efficient than electric resistance models because the electricity is used for moving heat from one place to another rather than for generating the heat directly (see discussion on heat pumps in Chapters 4 and 5). The heat source is the outside air or air in the basement where the unit is located. Refrigerant fluid and compressors are used to transfer heat into an insulated storage tank. While the efficiency is higher, so is the cost to purchase and maintain these units. Heat pump

Heat pump water heater

water heaters are available with built-in water tanks called integral units, or as add-ons to existing hot water tanks. A heat pump water heater uses one-third to one-half as much electricity as a conventional electric resistance water heater. In warm climates they may do even better.

Tankless Coil Water Heaters

Tankless coil water heaters use the home's main heating system as the heat source for water heating. They are common in older oil-fired boilers, although gas-fired boilers are also sometimes fitted with tankless coil water heaters. They operate directly off the house boiler; there is no separate storage tank. When hot water is drawn from the tap, the water circulates through a heat exchanger in the boiler. As long as the boiler is being used regularly (during the winter months), a tankless coil works very well because the boiler is usually hot. During the summer, spring, and fall months, however, the boiler has to cycle on and off frequently just for water heating, wasting energy the same way your car wastes gas in

Tankless coil water heater.

stop-and-go traffic. Tankless coils can consume 3 Btu of fuel for every

hot water they provide, making them more expensive to operate than most other types of water heaters. In general, tankless coil water heaters are not recommended. There are other integrated space and water heating systems that are far more effective.

Indirect Water Heaters

Indirect water heater.

Indirect water heaters use the home's boiler or furnace as the heat source, but in a very different way than a tankless coil. In boiler systems, hot water from the boiler is circulated through a heat exchanger in a separate insulated tank. In the less common furnace-based systems, water in a heat exchanger coil circulates through the furnace to be heated, then through the water storage tank. Since hot water is stored in an insulated storage tank, the boiler or funace does not have to turn on and off as frequently, improving its fuel economy. Electronic controls determine when water in the tank falls below a preset temperature and triggers the boiler or furnace to provide heat as long as needed. The more sophisticated of these systems rely on a heat purge cycle to circulate leftover heat remaining in the heat exchanger into the water storage tank after the boiler or furnace shuts down, thereby further improving overall system efficiency.

Indirect water heaters, when used in combination with new, high-efficiency boilers or furnaces, are usually the least expensive way to provide hot water (see **COMPARING THE TRUE COSTS OF WATER HEATERS** below). These systems can be purchased in an integrated form, incorporating the boiler or furnace and water heater with controls, or as separate components. Gas, oil, and propane-fired systems are available. Any form of hydronic space heating—hydronic baseboards, radiators, or radiant heat—can be provided by boiler systems. Table 6.3 provides contact information for manufacturers of indirect systems and components.

Advanced Heating Systems With Integrated Water Heaters

If you're building a new home or upgrading your heating system at the same time you're choosing a new water heater, you might consider a combination water heater and space heating system. These systems, also called dual integrated appliances, put water heating and space heating functions in one package. Space heating is provided via warm-air distribution.

Integrated gas heaters feature a powerful water heater, with space heating provided as the supplemental end use. Heated water from the water heater tank passes through a heat exchanger in a fan-coil unit to heat air. The fan blows this heated air into the ducts to heat the home. Many combination systems of this type are available at low initial costs, but space and water heating efficiency is often less than that of conventional systems. Models incorporating a high-efficiency condensing water heater, such as the American Water Heater Polaris and Lennox CompleteHeat, are exceptions. These models realize efficiency gains over traditional equipment.

Households using electric water heating and a heat pump for space conditioning can reduce water heating costs by installing a multi-function heat pump system. Fully-integrated, single-unit systems are available, or an existing heat pump and storage water heater can be retrofitted with a specially-designed add-on heat pump water heater module. Multi-function air- and ground-source heat pump systems are available.

As you may have guessed, proper sizing of a dual integrated appliance is very important for economical performance, since both space heating and water heating are given from one "box." The manufacturers listed in Table 6.3 should be able to identify your local distributors and contractors who are familiar with these products and their installation.

Solar Water Heaters

As the name implies, these use energy from the sun to heat water. While the initial cost of a solar water heater is high, it can save a lot of money over the long term (see **COMPARING THE TRUE COSTS OF WATER HEATING** below). Solar water heaters are much less common than they were during the 1970s and early 1980s when they were supported by tax credits, but the units available today tend to be considerably less expensive and more reliable. For example, in New England, it is now possible to have a complete system that will provide approximately two-thirds of a typical family's hot water requirements installed for less than $3,000. At that price, solar water heaters compete very well with electric and propane water heaters on a life-cycle cost basis, though they are still usually more expensive than natural gas.

Solar water heaters are designed to serve as preheaters for conventional storage or demand water heaters (either gas or electric).

They work quite well with demand water heaters that have modulating temperature controls. Because the solar system preheats the water, the extra temperature boost required by the demand water heater is relatively low, and a high flow rate can be achieved.

If you have extra money to invest and want to do more for the environment, a solar water heater can be a good choice. But make sure you find a qualified installer who can properly design and size the back-up water heating system. Solar water heaters can be particularly effective if they are designed for three-season use, with your heating system providing hot water during the winter months. If you

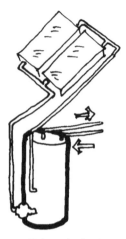

Solar water heater.

heat with wood, the solar water heater installer may be able to fit your wood stove with a water heating coil to take over in the winter months.

This edition of the *Consumer Guide* does not list individual solar water heating equipment; a list of manufacturers is provided later in Table 6.4.

SELECTING A NEW WATER HEATER

Whether you're replacing a worn-out existing water heater or looking for the best model for a new house you're building, it pays to choose carefully. Look for a water heater that satisfies your hot water needs and uses as little energy as possible. Often you can substantially reduce your hot water needs through water conservation efforts (see **Conserve Water** below).

Think About a Replacement Now

If you're like most people, you're unlikely to go out looking for a water heater until your existing one fails. That will happen at the worst possible time—like just after your in-laws arrive for a week-long visit. You'll have to rush out and put in whatever is available, without taking the time to look for a water heater that most appropriately fits your needs and offers the highest level of energy efficiency.

A much better approach is to do some research now. Figure out what type of water heater you want—gas or electric, storage or demand, stand-alone or integrated with your heating system, etc. Figure out the proper size for your household (not just gallon capacity, but first-hour rating as well—see **Sizing a Water Heater** below). And then use the listings here to identify your most energy-efficient options.

If possible, replace your existing water heater before it fails. Most water heaters have a lifespan of about 10-15 years. If yours is up there in age, have your plumber take a look at it and advise you on how much useful life it has left. If it's in bad shape, replace it now before it starts leaking or the burner stops working. In fact, it often makes sense to replace an inefficient water heater even if it's in good shape. The energy savings alone could pay for the new water heater after just a few years, and you'll be happy knowing that you are dumping fewer pollutants into the air and less money down the drain.

Sizing a Water Heater

To determine how big a storage water heater you need, you should first estimate your family's peak-hour demand. To do this, estimate what time of day (morning or evening) your family is likely to require the greatest amount of hot water. Then calculate the maximum expected hot water demand using Table 6.1. Note that this does not provide an estimate of your family's total daily use, only the peak hourly use. Also, the values in this table do not consider water conservation measures, like low-flow showerheads and faucet aerators, that can reduce hot water use for each activity.

The ability of a water heater to meet peak demands for hot water is indicated by its first-hour rating. This rating accounts for the effects of tank size and how quickly cold water is heated. In some cases, a water heater with a small tank but powerful burner can have a higher first-hour rating than one with a large tank and less powerful burner. First-hour ratings are provided in the listings that follow, along with tank size and efficiency (given as estimated energy use). You can also ask appliance dealers for the first-hour ratings of appliances they sell.

Buying too large a storage water heater will reduce energy performance by increasing the standby losses. If gas- or oil-fired, larger systems will also lose more heat up the flue.

Demand water heaters should be sized according to the required gpm flow rate and temperature rise required for your largest expected hot water fixture (usually a shower). (See discussion on demand water heaters above.)

With solar water heaters, you should discuss your requirements carefully with the solar water heating salesperson. You will need to size both the solar hot water system itself and the back-up electric or gas water heater. It generally makes the most sense to size a solar water heater to provide two-thirds to three-fourths of your total demand, and provide the rest with a back-up system.

Fuel Options

What type of fuel makes the most sense for your water heater? If you currently have an electric water heater and natural gas is available in your area, a switch might save you a lot of money. Oil- and propane-fired water heaters are also usually less expensive to operate than electric models.

Before you rule out electricity, though, check with your utility company. It may offer special off-peak rates that make electricity a more attractive option. With off-peak electricity for water heating, the utility company puts in a separate meter with a timer in it. You can only draw electricity through that meter during off-peak periods, when the utility company has more capacity than it needs and is willing to sell it less expensively.

TABLE 6.1

PEAK HOURLY HOT WATER DEMAND

	Avg. gallons hot water per usage		Times used in hour		Gal. used in hour
Showering	20	×	_____	=	_____
Bathing	20	×	_____	=	_____
Shaving	2	×	_____	=	_____
Washing hands and face	2	×	_____	=	_____
Shampooing hair	4	×	_____	=	_____
Hand dishwashing	4	×	_____	=	_____
Automatic dishwashing	12	×	_____	=	_____
Preparing food	5	×	_____	=	_____
Automatic clothes washing	32	×	_____	=	_____

For example, if your family's expected greatest hot water use is in the morning, the total might be:

3 showers	20	×	3	=	60 gallons/hr.
1 shave	2	×	1	=	2 gallons/hr.
Hand-wash dishes	4	×	1	=	4 gallons/hr.
Peak hour demand				=	66 gallons/hr.

Source: Gas Appliance Manufacturers Association and ACEEE estimates.

Look for Sealed Combustion or Power-Vented Systems

For safety as well as energy efficiency reasons with gas- and oil-fired water heaters, look for units with sealed combustion or power venting. Sealed combustion means that outside air is brought in directly to the water heater and exhaust gases are vented directly outside. The combustion is totally separated from the house air. Power-vented equipment can use house air for combustion, but flue gases are vented to the outside with the aid of a fan.

In very tight houses, drawing combustion air from the house and passively venting flue gases up the chimney can sometimes result in back-drafting of dangerous combustion gases into the house.

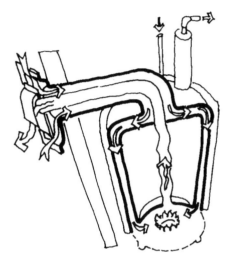

Sealed combustion gas water heater.

WATER HEATER EFFICIENCY

The energy efficiency of a storage water heater is indicated by its *energy factor* (EF), an overall efficiency based on the use of 64 gallons of hot water per day. The national appliance efficiency standards for water heaters that took effect in 1990 require the energy factors listed below, based on storage tank size:

MINIMUM ENERGY FACTORS (EF)

Tank size	Gas	Oil	Electric
30 gallons	0.56	0.53	0.91
40 gallons	0.54	0.53	0.90
50 gallons	0.53	0.50	0.88
60 gallons	0.51	0.48	0.87

All other things being equal, the smaller the water heater tank, the higher the efficiency. Compared to small tanks, large tanks have a greater surface area, which increases heat loss from the tank and decreases the energy efficiency somewhat, as mentioned above. If your utility company offers off-peak electric rates and you'd like to use them, you may need to buy a larger water heater to provide carry-over hot water for periods when electricity is not available.

THE MOST EFFICIENT
WATER HEATERS

The tables below list the most efficient water heaters on the market today. For storage water heaters, the first-hour ratings, storage capacities, and estimated annual energy use are listed. The most efficient electric storage water heaters all have energy factors between 0.94 and 0.96, resulting in estimated annual energy use below 4,675 kWh/year. Note that heat pump water heaters use less than half as much electricity as conventional electric resistance water heaters. The most efficient gas-fired storage water heaters have energy factors ranging from 0.62 to 0.68, corresponding to estimated gas use below 250 therms/year. Condensing water heaters have energy factors as high as 0.86.

The efficiency of an integrated system is given by its combined annual efficiency, which is based on the AFUE of the space heating component and the energy factor of the water heating component. We list the most efficient combination units, only including those with combined annual efficiencies of 0.75 (75%) or higher.

If your local plumber or hardware store doesn't carry a model listed here, call the manufacturer (see Appendix 3 for phone numbers).

MOST EFFICIENT ELECTRIC WATER HEATERS

Manufacturer	Model	1st Hour Rating (gallons)	Storage Tank (gallons)	kWh/yr
Approximately 30 gallon storage tank				
Sears	449.310311	39	30	4,624
Marathon	MP30245	42	30	4,624
Sears	449.320311	46	30	4,624
Maytag	HE*X-30T-961	42	30	4,671
Sears	153.32035* HT	42	30	4,671
Marathon	**30238 A	39	30	4,671
Reliance	** 30 2LRT*** W	45	30	4,671

MOST EFFICIENT ELECTRIC WATER HEATERS (cont.)

Manufacturer	Model	1st Hour Rating (gallons)	Storage Tank (gallons)	kWh/yr
Approximately 30 gallon storage tank (cont.)				
Reliance	** 30 2ORS973 W	45	30	4,671
State	CD** 30 2ORS*** W	45	30	4,671
State	PEX 30 2PRT*** W	45	30	4,671
State	SSX 30 2LR**** W	45	30	4,671
Marathon	MR30245 A	42	30	4,671
Approximately 40 gallon storage tank				
Sears	449.314410	45	40	4,624
Sears	449.310411	49	40	4,624
Marathon	MP40245	52	40	4,624
Sears	449.320411	56	40	4,624
Maytag	HE*X-40S-961	51	40	4,671
Marathon	**40238 A	48	40	4,671
Sears	449.311410	48	40	4,671
Marathon	MR40245 A	52	40	4,671
Approximately 50 gallon storage tank				
Marathon	MX50230S	47	50	4,624
Marathon	MX50245S	54	50	4,624
Marathon	MSR50238 A	51	50	4,671
Marathon	MSR50245 A	54	50	4,671
A.O. Smith	*EH-52	60	50	4,671
Marathon	** 50238 A	58	50	4,671
Marathon	MR50245 A	61	50	4,671
Sears	449.311510	58	52	4,671
Sears	449.310531	59	52	4,671
Sears	449.311530	66	52	4,671
Sears	449.320511	66	52	4,671
Approximately 60 gallon storage tank or larger				
Marathon	MX75238	75	75	4,624
Marathon	MX75245	78	75	4,624
Marathon	MX85238	87	85	4,624
Marathon	MX85245	90	85	4,624
Marathon	MX85255	94	85	4,624

MOST EFFICIENT ELECTRIC HEAT PUMP WATER HEATERS

Brand	Model	1st Hour Rating (gallons)	Storage Tank (gallons)	kWh/yr
DEC/Therma-Stor	TS-VHP-80	64	80	1,753
DEC/Therma-Stor	TS-HP-120-18-30	99	120	1,753
Crispaire/E-Tech	WH6-BX-1	45	add-on	2,000
Crispaire/E-Tech	R106K3	58	add-on	2,306

MOST EFFICIENT GAS WATER HEATERS

Manufacturer	Model	1st Hour Rating (gallons)	Storage Tank (gallons)	Therms
Approximately 30 gallon storage tank				
American Water Heater Co.	PC100-35	147	34	174
American Water Heater Co.	PR35	147	34	174
Bock	32PPPG	131	32	221
Bock	30ESPG	109	28	238
A. O. Smith	BGC-30T-242	64	30	241
A. O. Smith	FGR-30-240	64	30	241
A. O. Smith	FGR-30T-242	64	30	241
Approximately 40 gallon storage tank				
A.O. Smith	F*SE-40-230E	75	40	231
A.O. Smith	FPCR-40-234	75	40	231
Bradford-White	M-4-403T***N-11	73	38	234
Bradford-White	M-4-403T***N-13	79	40	234
Bradford-White	M-4-40T***N-13	67	40	234
Bradford-White	M-I-TW40L***N-11	70	38	234
Maytag	HN*X-40-X-960	76	40	234
A.O. Smith	PGCG-40-226	76	40	238
A.O. Smith	PGCG-40-246	81	40	238
Bradford-White	M-4-403T***N-10	73	38	238
Bradford-White	M-4-403T***N-12	79	40	238
Bradford-White	M-4-40T***N-12	67	40	238
Bradford-White	M-I-TW40L***N-10	70	38	238
Lochinvar	DVN041	67	38	238
Reliance	** 40 N*CT*** 52W	73	40	238
Reliance	** 40 NXRTL*** W	76	40	238
Rheem	44V40-1	71	39	238
Rheem	44V40-1N	71	39	238
Sears	153.33040* HA	76	40	238
Sears	153.33045*	76	40	238
State	PR* 40 NOCT*** 52W	73	40	238
Approximately 50 gallon storage tank				
American Water Heater Co.	PBG102-50T100-2NV	147	50	174
American Water Heater Co.	PBG*2-50T100-2NV	147	50	174
American Water Heater Co.	PC100-50	147	50	174
American Water Heater Co.	PR50	147	50	174
A.O. Smith	FPCR-50-234	82	50	231
A.O. Smith	F*SE-50-230E	84	50	231
Bradford-White	M-II-TW50T***N-10	105	48	231
Bradford-White	M-I-TW50L***N-11	85	48	234
Maytag	HN*X-50X-960	84	50	234
Apollo	A5 50 40.ONXRT*** 2W	84	50	238
Bradford-White	M-4-50S***N-13	85	50	238
Bradford-White	M-I-TW50L***N-10	85	48	238
Lochinvar	DVN051	80	48	238

MOST EFFICIENT GAS WATER HEATERS (cont.)

Manufacturer	Model	1st Hour Rating (gallons)	Storage Tank (gallons)	Therms
Approximately 50 gallon storage tank (cont.)				
Reliance	** 50 NXRT*** 2W	84	50	238
Reliance	** 50 NXRTL*** 2W	84	50	238
State	PR* 50 NXRT*** 2W	84	50	238

MOST EFFICIENT GAS COMBINATION WATER HEATERS/SPACE HEATERS

Manufacturer	Model	Heating Capacity (Btu/hr)	Combined Annual Efficiency
Lennox/Complete Heat	HM30-100	90,000	0.90
Lennox/Complete Heat	HM30-150	135,000	0.90
Polaris/Amer. Wtr. Htr.	PBG102-34S100-2N	90,000	0.90
Polaris/Amer. Wtr. Htr.	PBG102-50T100-2N	89,000	0.88

MOST EFFICIENT OIL WATER HEATERS

Manufacturer	Model	1st Hour Rating (gallons)	Storage Tank (gallons)	Oil Use (gal/yr)
Bock	32PP**	131	32	159
Bock	32E**	134	32	164
Bock	30ES**	109	28	169
Heat Transfer	SJ-30 CWB	101	30	175
GSW Water Heating	JW 3*7	150	30	175
GSW Water Heating	JWF 307	110	32	180
GSW Water Heating	JWF 307V	110	32	180
GSW Water Heating	JWF 307H	150	32	186
GSW Water Heating	JWF 507	190	50	196
Bradford-White	M-I-32L**OF-10	120	32	200
Bradford-White	M-I-50L**OF-10	140	50	213

DEMAND OR "INSTANTANEOUS" WATER HEATERS

With demand water heaters, different specifications are provided: the energy input (Btu/hour for gas, kilowatts [kW] for electric); the temperature rise achievable at the rated flow; the flow rate at the listed temperature rise; the minimum flow rate required to fire the heating elements; whether or not the unit is available with a modulating temperature control to automatically adjust the amount of heating required to maintain the desired delivery temperature; and the maximum water pressure. In comparing the different models, be aware that you aren't always looking at direct comparisons, especially with temperature rise and flow rate. For example, while one model might list the flow rate at a 100°F temperature rise, another might list the flow rate at 70°. Until there are industry-standard ratings for temperature rise and flow rates, it will be difficult to compare the performance of products from different companies.

INSTANTANEOUS WATER HEATERS – ELECTRIC

Brand	Model	Low/high kW	Temp. change	Flow rate	Min. flow	Modulation	Min/max
Powerstream	RP1/240	4.75/9.5	64	1.0	0.75	n	15/140
Powerstream	RP1/208	3.5/7.1	48	1.0	0.75	n	15/150
Powerstream	RP2	3/6	40	1.0	0.5	n	15/150
Powerstream	RP3	3	20	1.0	0.5	n	15/150

INSTANTANEOUS WATER HEATERS – GAS

Brand	Model	Low/high 1000 Btu/hr	Temp. change	Flow rate	Variable min. flow	Input burner	Min/max
Aquastar	38	38.7	90	0.8	0.75	y	15/150
Aquastar	80	25/77.5	90	1.3	1.1	y	15/150
Aquastar	125	25/125	90	2.1	0.75	y	15/150
Aquastar	170	56/165	90	3.0	0.75	y	15/150
Myson	325	20/100	90	1.7	0.75	y	35/150
Paloma	PH-5-3F	38.1	100	0.6	0.5	n	Min 4.3
Paloma	PH-6D	43.8	100	0.7	0.5	n	Min 4.3
Paloma	PH-12MD	30/89.3	100	1.4	0.5	y	Min 2.1
Paloma	PH-24MD	37.7/178.5	100	2.9	0.5	y	Min 2.4
Vaillant	MAG 325	43/100	90	1.7	n/a	y	n/a

TABLE 6.2

DISTRIBUTORS OF INSTANTANEOUS ("TANKLESS") WATER HEATERS

Brands: AquaStar, Ariston, Powerstream	Controlled Energy Corporation Fiddler's Green Waitsfield, VT 05673 800-642-3111 802-495-4436 www.cechot.com
Brand: Paloma	Low Energy Systems, Inc. 2916 S. Fox St. Englewood, CO 80110 800-873-3507 303-781-9437 www.palomawaterheaters.com
Brand: Myson	McNeely-Yuill Corp. 9911 Horn Road, Suite 100-A Sacramento, CA 95287 800-456-3761 916-362-1671 www.plumbingnet.com/myson

TABLE 6.3

MANUFACTURERS OF COMBINATION WATER/SPACE HEATING SYSTEMS

For gas service:
American Water Heater (Polaris)
P.O. Box 1597
Johnson City, TN 37605
800-456-9805
www.americanwaterheater.com

Apollo Comfort Products
500 Bypass Road
Ashland City, TN 37015
800-365-8170
www.stateind.com

Lennox (CompleteHeat)
P.O. Box 799900
Dallas, TX 75379-9900
800-953-6669
www.lennox.com

Trianco-Heatmaker
20 Industrial Way
P.O. Box 1000
Rochester, NH 03867
www.teledynelaars.com

For oil service:
Energy Kinetics
51 Molasses Hill Road
Lebanon, NJ 08833
800-323-2066
www.energykinetics.com

Indirect water heaters:
Ergomax
10 Summit Avenue
Berkeley Heights, NJ 07922
908-665-0700
www.ergomax.com

TABLE 6.4

DOMESTIC SOLAR WATER HEATING SYSTEMS

American Energy Technologies
1057 Ellis Rd, Unit 4
Jacksonville, FL 32254
800-874-2190   904-781-7000
www.aetsolar.com

Solar Development Incorporated
3607 A Prospect Ave.
Riviera Beach, FL 33404
561-842-8935
www.solardev.com

Radco
2877 Industrial Parkway
Santa Maria, CA 93455
800-927-2326

Thermo Dynamics, Ltd.
44 Borden Ave.
Dartmouth, Nova Scotia B3B 1C8
Canada
902-468-1001

COMPARING THE TRUE COSTS OF WATER HEATERS

There are a number of important considerations when deciding what type of water heater you should buy: fuel type, efficiency, configuration (storage, demand, integrated), size, and cost. The information above covers most of these issues. Cost, however, needs some additional discussion. There are really two types of cost you need to look at: purchase cost and operating cost.

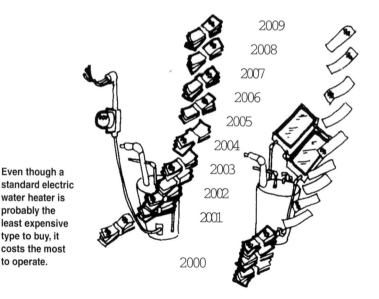

2009
2008
2007
2006
2005
2004
2003
2002
2001
2000

Even though a standard electric water heater is probably the least expensive type to buy, it costs the most to operate.

It may be tempting when you're buying a water heater simply to look for a model that is inexpensive to buy, and ignore the operating cost. This is a poor strategy. Often, the least expensive water heaters to buy are the most expensive to operate. Life-cycle costs, which take into account both the initial costs and operating costs of different water heaters, provide a much more accurate representation of the true costs of the water heater. Life-cycle costs for the most common types of water heaters under typical operating conditions are shown in Table 6.5.

From the table, we see that when both purchase and operating costs are taken into account, one of the least expensive systems to buy (conventional electric storage) is one of the most costly to operate over a 13-year period. An electric heat pump water heater, though expensive to purchase, has a much lower cost over the long term. A solar water heating system, which costs the most to buy, has the lowest yearly operating cost.

TABLE 6.5

LIFE-CYCLE COSTS FOR 13-YEAR OPERATION OF DIFFERENT TYPES OF WATER HEATERS

Water heater type	Efficiency	Cost[1]	Yearly energy cost[2]	Life (years)	Cost over 13 years[3]
Conventional gas storage	55%	$425	$163	13	$2,544
High-eff. gas storage	62%	$500	$145	13	$2,385
Oil-fired free-standing	55%	$1,100	$228	8	$4,751
Conventional electric storage	90%	$425	$390	13	$5,495
High-eff. electric storage	94%	$500	$374	13	$5,362
Demand gas	70%	$650	$140	20	$2,243
Demand electric (2 units)	100%	$600	$400	20	$5,590
Electric heat pump	220%	$1,200	$160	13	$3,280
Indirect water heater with efficient gas or oil boiler	75%	$700	$150	30	$2,253
Solar with electric back-up	n/a	$2,500	$125	20	$3,250

1. Approximate. Includes installation.
2. Energy costs based on hot water needs for typical family of four and energy costs of 8¢/kWh for electricity, 60¢/therm for gas, $1.00/gallon for oil.
3. Future operation costs are neigher discounted nor adjusted for inflation.
Source: American Council for an Energy-Efficient Economy.

INSTALLING A WATER HEATER

Select an installation contractor carefully. Make sure that he or she has experience with the type of system you want to put in. If the system is integrated with your heating system, have your heating contractor put in the water heater.

To get a good price, ask for bids from several contractors and evaluate the bids carefully. Consider warranties, service, and reputation as well as the price.

Storage water heaters will lose less heat if they are located in a relatively warm area. Don't install the water heater in an unheated basement if at all possible. Also try to minimize the length of piping runs to your kitchen and bathrooms. The best location is a centralized one, not too far from any of your hot water taps.

When the water heater is being installed, make sure heat traps or one-way valves are installed on both the hot and cold water lines to cut down on losses through the pipes. Without heat traps, hot water rises and cold water falls within the pipes, allowing heat from the water heater to be lost to the surroundings. Heat traps cost around $30 and will save $15-30 per year. Some new water heaters have built-in heat traps. If heat traps are not installed, you should insulate several feet of cold water pipe closest to the water heater in addition to the hot water pipes. Even with heat traps, insulate the cold water line between the water tank and heat trap.

Also follow general energy conservation tips when you are installing a new water heater (see **UPGRADING YOUR EXISTING WATER HEATER** below). Just because it's a new water heater, for example, doesn't mean it won't benefit from an added insulation blanket.

UPGRADING YOUR EXISTING WATER HEATER

Even if you aren't going to buy a new water heater, you can save a lot of energy and money with your existing system by following a few simple suggestions.

Conserve Water

Your biggest opportunity for savings is to use less hot water. In addition to saving energy (and money), cutting down on hot water use helps conserve dwindling water supplies, which in some parts of the country is a critical problem. A family of four each showering five minutes a day can use about

Low-flow showerheads and faucet aerators reduce your hot water use, saving energy and money.

700 gallons per week—a three-year drinking water supply for one person! Water-conserving showerheads and faucet aerators can cut hot water use in half. That family of four can save 14,000 gallons of water a year and the energy required to heat it.

If you aren't sure how much water your shower currently uses, you can find out with a bucket that holds at least a gallon and with a watch with a second hand. Turn on the shower to the usual pressure you use, hold the bucket under it and time how many seconds it takes to fill to the one-gallon mark. If it takes less than 20 seconds, your flow rate is over 3 gpm, and your shower is a good candidate for a low-flow showerhead. Many older showerheads typically deliver 4–5 gpm.

In fact, federal efficiency standards now require that all new showerheads perform at no more than 2.5 gallons per minute. A variety of high-quality showerheads meet this standard while providing very satisfactory performance, including some of the shower massage products that have become more popular.

New faucets are also subject to the 2.5 gpm standard, but there are opportunities to save even more. Look for faucet aerators that deliver 1/2-1 gpm, especially if you tend to run the water when rinsing dishes. Some models are sold with a convenient shut-off valve at the aerator that allows you to temporarily turn off the water without changing the hot/cold mix. Sink aerators cost just a few dollars apiece. Repair leaky faucets right away. Even a relatively small hot water drip can dump a lot of energy and money down the drain.

Also, refer to Chapters 9 and 10 for water-saving tips with dishwashers and washing machines. In addition to saving money, any measure to conserve water will also reduce your water and sewer bills.

Insulate Your Existing Water Heater

Installing an insulating jacket on your existing water heater is one of the most effective do-it-yourself energy-saving projects. The insulating jacket will reduce standby heat loss—heat lost through the walls of the tank—by 25-40%, saving 4-9% on your water heating bills. Water heater insulation jackets are widely available for just $10-20

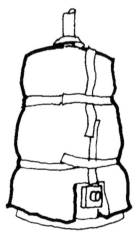

Even with new water heaters, it is usually a good idea to wrap them with insulation.

(sometimes less from a utility company). An insulating jacket will pay for itself through energy savings in less than a year. Add at least an additional R-8 to the sides and top of the tank.

Some newer water heaters come with fairly high insulation levels, reducing (though not eliminating) the economic advantages of adding additional insulation. In fact, some manufacturers recommend against installing insulating jackets on their energy-efficient models. Always follow directions carefully when installing an insulation jacket. Leave the thermostat(s) accessible. With gas- and oil-fired water heaters, you need to be careful not to restrict the air inlet(s).

Insulate Hot Water Pipes and Install Heat Traps

Insulating your hot water pipes will reduce losses as the hot water is flowing to your faucet and, more importantly, it will reduce standby losses when the tap is turned off and then back on within an hour or so. A great deal of energy and water is wasted waiting for the hot water to reach the tap. Even when pipes are insulated, the water in the pipes

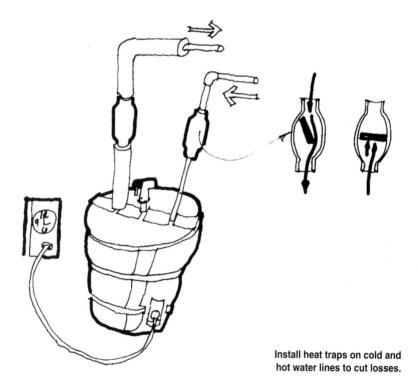

Install heat traps on cold and hot water lines to cut losses.

will eventually cool, but it stays warmer much longer than it would if the pipes weren't insulated.

Also have heat traps installed, as described above, if they are not already in place (see **INSTALLING A WATER HEATER** above).

Lower the Water Heater Temperature

Keep your water heater thermostat set at the lowest temperature that provides you with sufficient hot water. For most households, 120°F water is fine (about midway between the "low" and "medium" setting. Each 10°F reduction in water temperature will generally save 3-5% on your water heating costs. If you have a dishwasher without a booster heater, you should probably keep the water temperature up to 140°F (the "medium" setting) or buy a new, more efficient, dishwasher (see Chapter 9). Not only will lowering the thermostat save a lot of energy, but it will also increase the life of your water heater and re-

You will save a lot of energy and money by lowering the thermostat setting on your water heater.

duce the risk of children scalding themselves with the hot water.

Electric water heaters often have two thermostats—one for the upper heating element and one for the lower heating element. These should be adjusted to the same level to prevent one element from doing all the work and wearing out prematurely. Be sure to turn off the electricity at the circuit breaker before removing the access panels on an electric water heater.

When you are going away on vacation, you can turn the thermostat down to the lowest possible setting, or turn the water heater off altogether for additional savings. With a gas water heater, make sure you know how to relight the pilot if you're going to turn it off while away.

RECOMMENDATIONS

1. Plan ahead. Don't wait until your existing water heater fails before you look into replacement. It often makes very good economic sense to replace an old inefficient model with a high-efficiency one even before the old one fails.

2. Take into account life-cycle costs when you're choosing a water heater. Don't be tempted to simply buy the lowest price model available.

3. With storage water heaters, buy an efficient model, such as the ones listed in this chapter.

4. Implement water conservation strategies. You'll save energy and money whether you're buying a new system or not. Install low-flow showerheads and faucet aerators. Refer to Chapters 9 and 10 for water-saving tips with other appliances.

5. Insulate your water heater if it's a storage model. Even new energy-efficient models should generally be insulated.

6. Insulate hot water pipes.

7. Lower the temperature setting on your water heater to about 120°F.

CHAPTER 7
Food Storage

The energy use of refrigerators and freezers has improved dramatically in the past 20 years, but they are still among the largest energy consumers in the home. A typical new refrigerator today uses less than 700 kWh per year, while the typical model sold in 1973 used nearly three times as much. Moreover, the typical unit today is larger and has better controls. This increase in efficiency has been achieved through more insulation, tighter door seals, larger coil surface area, and improved compressors and motors.

Much of the increase in efficiency is due to national energy efficiency standards for new refrigerators. The current refrigerator standards took effect January 1, 1993, and new standards will take effect in 2001, lowering energy consumption even more. In some states, utilities offer rebates for the purchase of models that exceed the minimum standards.

BUYING A NEW REFRIGERATOR

When it comes time to buy a new refrigerator, it definitely pays to shop around for an energy-efficient model. Even though federal law mandates certain energy efficiency levels for refrigerators, there is still significant variation from model to model. Look for the ENERGY STAR® label to identify efficient models, and use the listings below for guidance.

As you shop for a new refrigerator, consider what style and features you want, and what the energy penalties might be. For example, side-by-side refrigerator/freezers use more energy than similarly sized models with the freezer on the top. Built-in designer refrigerators may also consume more energy than store models, but are less wasteful than they used to be since the national appliance energy standards took effect. Manual defrost models use less electricity than automatic defrost models but are not widely available in large sizes. However, manual defrost models must be defrosted periodically to maintain their energy efficiency. Features such as automatic icemakers and through-the-door dispensers can increase energy consumption somewhat.

Consider size as well when shopping for a refrigerator. Generally, the larger the unit, the greater the energy consumption. Too large a model will result in wasted space and energy; too small a model could mean extra trips to the supermarket. However, some refrigerator sizes tend to be more efficient. Currently, the most efficient models are in the most popular 16-20 ft³ range. You may find that a more efficient 18 ft3 model costs less to run than a 15 ft³ model with similar features.

If you are thinking of buying a second refrigerator, you might want to reconsider. It is generally much less expensive to buy and operate one big refrigerator than two small ones. If the extra refrigerator is an old model, it's probably an energy guzzler. If you only need a second refrigerator a few days a year or to hold a few six-packs of beer, why spend an extra $50-150 per year in electricity?

The Most Efficient Refrigerators

The most energy-efficient refrigerator and refrigerator/freezer models are listed below. The listings are organized according to configuration and size. Models are listed within each group in order of increasing electricity use (although technically, a larger refrigerator that uses only slightly more electricity than a smaller model may be considered more efficient on the basis of cubic feet per kWh consumed).

All of the refrigerators listed below are at least 20% more efficient than the minimum efficiency required by law. Check with your utility to see if a rebate is available for a high-efficiency model, and look for the ENERGY STAR® label, too.

MOST EFFICIENT REFRIGERATORS				
Brand	Model	Volume	Energy Use (kWh/yr)	Annual Energy Cost ($)
Top freezer, automatic defrost, 14.3 - 16.4 cubic feet				
Magic Chef	CT*1511*EW	15.0	437	37
General Electric	TBH14*AT	14.4	496	42
Roper	RT14HB*F*0*	14.3	498	42
Roper	RT14HD*G*0*	14.3	498	42
Roper	RT14HD*D*0*	14.3	498	42
Roper	RT14HD*E*0*	14.3	498	42
Roper	RT14WK*G*0*	14.3	498	42
General Electric	TBH14DAX	14.4	499	42
Hotpoint	CTH14CY*	14.4	499	42
General Electric	TBH14DAZ	14.4	500	42
General Electric	TBH16JA*	15.5	514	43
Hotpoint	CTH16*Y*	15.6	514	43
Top freezer, automatic defrost, 16.5 - 18.4 cubic feet				
Whirlpool	ET18HP*H*0*	18.2	514	43

MOST EFFICIENT REFRIGERATORS (cont.)

Brand	Model	Volume	Energy Use (kWh/yr)	Annual Energy Cost ($)
Top freezer, automatic defrost, 16.5 - 18.4 cubic feet (cont.)				
General Electric	TBH18JAZ	18.2	518	44
Roper	RT18HD*D*0	17.9	551	46
Roper	RT18HD*H*0*	18.2	554	47
General Electric	TBH18*AX	18.2	555	47
Top freezer, automatic defrost, 18.5 - 20.4 cubic feet				
Jenn-Air	JTB1988DE*	18.5	485	41
Maytag	MTB1956DE*	18.5	485	41
Jenn-Air	JTB1988AE*	18.5	527	44
Maytag	MTB1946*E*	18.5	527	44
Maytag	MTB1956AE*	18.5	527	44
Maytag	MTB1956BE*	18.5	527	44
Maytag	MTF1956AE*	18.5	527	44
General Electric	TBH19ZAZ	19.0	533	45
Kitchen Aid	KTRS19KH**0*	18.8	560	47
Whirlpool	ET19DK*E*0*	18.8	560	47
Whirlpool	GT19DK*H*0*	18.8	560	47
Kitchen Aid	KTRS20KH**0*	20.4	595	50
Kitchen Aid	KTRS20MH**0*	20.4	595	50
Top freezer, automatic defrost, 20.5 - 22.4 cubic feet				
Jenn-Air	JTB2188DE*	20.7	515	43
Maytag	MTB2156DE*	20.7	515	43
Maytag	MTF2156AE*	20.7	559	47
Jenn-Air	JTB2188AE*	20.7	559	47
Maytag	MTB2156AE*	20.7	559	47
Maytag	MTB2156BE*	20.7	559	47
Kitchen Aid	KTRS22MH**0*	21.6	611	51
Top freezer, automatic defrost, 22.5 cubic feet and larger				
Jenn-Air	JTB2488DE*	23.7	560	47
Maytag	MTB2456DE*	23.7	560	47
Maytag	MTF2456DE*	23.7	560	47
Jenn-Air	JTB2488AE*	23.7	598	50
Maytag	MTB2446*E*	23.7	598	50
Maytag	MTB2455BRW	23.7	598	50
Maytag	MTB2456AE*	23.7	598	50
Jenn-Air	JTB2688*E*	25.5	620	52
Jenn-Air	JTF2688*E*	25.5	620	52
Jenn-Air	JTF2688ARW	25.5	620	52
Maytag	MTB2656*E*	25.5	620	52
Maytag	MTF2656DE*	25.5	620	52
Maytag	MTF2656AE*	23.7	620	52
Amana	T*25*	24.5	649	55

MOST EFFICIENT REFRIGERATORS (cont.)

Brand	Model	Volume	Energy Use (kWh/yr)	Annual Energy Cost ($)
Top freezer, automatic defrost, 22.5 cubic feet and larger (cont.)				
Amana	TRI25V	24.5	649	55
Amana	TSI25T	24.5	649	55
Amana	TSI25V	24.5	650	55
Bottom freezer, automatic defrost				
Amana	B*21V	20.5	576	48
Amana	B*I21V	20.5	576	48
Modern Maid	MYB21A	20.5	576	48
Amana	BBI20T**	19.7	593	50
Amana	BRF20T	19.7	593	50
Amana	BR22V	21.7	594	50
General Electric	TCX22ZPAC	21.7	594	50
Kenmore	596.**27**	21.7	594	50
Kenmore	*727R	21.7	594	50
Kitchen Aid	KBRS22KG***1	21.8	634	53
Side-by-side, 19.5 - 22.4 cubic feet, with through-the-door ice				
Kitchen Aid	KSRE22FH**0*	21.6	665	56
Whirlpool	GD22YF*H*0*	21.7	665	56
Kenmore	5929*99*	21.6	714	60
Kitchen Aid	KSR*22F***0*	21.6	714	60
Kitchen Aid	KSR*22Q***0*	21.6	714	60
Whirlpool	ED22DQ***0	21.7	714	60
Whirlpool	GD22D**F*0	21.7	714	60
Whirlpool	GD22SF*H*0	21.7	714	60
Jenn-Air	JCD2289A**	22.2	720	60
Whirlpool	ED20D*XE*0*	19.7	734	62
General Electric	TFH22JWT	21.7	756	64
Kenmore	95727*	21.7	759	64
Whirlpool	ED22PQ*D*0	21.7	760	64
Side-by-side, really big ones, with through-the-door ice				
Jenn-Air	JSD2388AE*	22.9	685	58
Maytag	MSD2356AE*	22.9	685	58
Kitchen Aid	KSRS25QD**0*	25.1	724	61
Whirlpool	GD25YF*H*0*	25.2	724	61
Kitchen Aid	KSRE25FH**0*	25.1	725	61
Jenn-Air	JSD2789AT*	26.7	735	62
Jenn-Air	JSD2789AE*	26.6	735	62
Maytag	MSD2756AE*	26.5	735	62
Maytag	MSD2757AE*	26.6	735	62
Kitchen Aid	KSRE27FH**0*	26.5	744	62
Whirlpool	GD27YF*H*0*	26.6	746	63
Jenn-Air	JSD2588AE*	24.7	760	64

Brand	Model	Volume	Energy Use (kWh/yr)	Annual Energy Cost ($)
MOST EFFICIENT REFRIGERATORS (cont.)				
Side-by-side, really big ones, with through-the-door ice (cont.)				
Maytag	MSD2556AE*	24.7	760	64
Jenn-Air	JSD2989AE*	28.6	765	64
Maytag	MSD2957AE*	28.6	765	64
Kenmore	5959*99*	25.2	777	65
Kitchen Aid	KSR**25F***0*	25.1	777	65
Whirlpool	GD25*F***0*	25.2	777	65
Whirlpool	GD25DQ*F*0*	25.2	777	65
General Electric	TFH24JRS	23.6	796	67
Kitchen Aid	KS**27Q***0*	26.5	798	67
Kitchen Aid	KSR*27F***0*	26.5	798	67
Whirlpool	GD27DF*F*0*	26.6	798	67
Whirlpool	GD27DQ*F*0*	26.6	798	67
Roper	RS25AW*E*0	25.2	825	69
Whirlpool	ED25TQ*E*0*	25.2	825	69
Maytag	MSD2354AR*	22.8	830	70
General Electric	TFH27PRT	26.6	851	71
Kitchen Aid	KSRS27QA**0*	26.5	851	71
Whirlpool	ED27PQ*E*0	26.6	851	71
Maytag	MSD2554*R*	24.7	852	72

BUYING A NEW FREEZER

There are two basic freezer styles: upright (front loading) and chest (top loading). Chest freezers are 10-25% more efficient than uprights because they are better insulated and air doesn't spill out when the door is opened. However, you should also consider convenience when selecting a freezer—chest models can be more difficult to organize.

Manual defrost freezers are more common than automatic defrost models, and they tend to do a better job at storing food. (Automatic defrost freezers may dehydrate frozen foods, causing freezer burn.) Because a freezer is opened less frequently than a refrigerator, frost will generally not build up as quickly as it might in manual defrost refrigerators.

The Most Efficient Freezers

New efficiency standards took effect in 1993 for freezers, too. At this time there are no significant differences in efficiency between different brands of new freezers of similar size and style—all are much more efficient than older models.

The following chart shows typical annual electricity use (and cost) for new freezers, by configuration and common sizes.

Size	Chest freezer		Upright freezer	
	kWh/yr	$/yr	kWh/yr	$/yr
7 cubic feet	290	25	400	34
12 cubic feet	380	32	470	40
15 cubic feet	425	36	525	45
20 cubic feet	510	43	575	49

INSTALLATION

If possible, locate refrigerators and freezers away from heat sources and direct sunlight. In the kitchen, try to keep your refrigerator away from the dishwasher and oven (keep it on a short leash, if necessary). Allow at least 1" space on each side of a refrigerator or freezer to allow good air circulation. Freezers can be installed in an attached garage or basement, which will boost energy performance somewhat during cooler months and reduce cooling loads in your house somewhat during the warmer months. However, don't put a refrigerator or freezer in a space that frequently goes below 45°—the refrigerant will not work properly.

BOOSTING THE EFFICIENCY
OF YOUR EXISTING REFRIGERATOR OR FREEZER

From an energy standpoint, you will save the most by replacing your existing refrigerator or freezer with a new, more efficient model. If it is more than about 15 years old, it may be so inefficient that a new one would pay for itself in energy savings in just a few years. Unfortunately, that is often not practical. Refrigerators cost a lot. If your present one is working fine, it's hard to justify running out to buy a new one. So here are a number of ways to boost energy efficiency and performance of refrigerators and freezers.

Check Door Seals

Check the door seals or gaskets on your refrigerator/freezer. These can deteriorate over time, greatly increasing heat gain and decreasing energy performance. Put a dollar bill in the door as you close it; if it is not held firmly in place, the seals are probably defective. With newer magnetic door seals, this test may not work. Instead, put a bright 150-watt flood lamp inside the refrigerator and direct the light toward a section of the door seal. With the door closed and the room light dimmed, inspect for light through the crack. You will have to reposition the light as you move along the perimeter of the seal. Use a mirror to check the seal at the bottom of the door. If you don't see light, the seals should be in good shape.

The dealer you purchased the refrigerator or freezer from should be able to install new seals. New seals aren't cheap, though. If the seals are bad, you might want to evaluate whether it's time to buy a new, high-efficiency model.

Inspect your refrigerator door gaskets for tightness with a piece of paper or a light.

Check the Temperature

Check the temperature inside your refrigerator and freezer with an accurate thermometer. The refrigerator compartment should be kept between 36°F and 38°F, and the freezer compartment between 0°F and 5°F. If the temperature is outside these ranges, adjust the thermostat control. Keeping temperatures 10°F lower than these recommended levels can increase energy use by as much as 25%.

Move the Refrigerator to a Cooler Location

Take a look at where the refrigerator is located. If it's in the sunlight or next to your stove or dishwasher, it has to work harder to maintain cool temperatures. If you can move it to a cooler location, you'll boost energy performance. Also, make sure that air can freely circulate around the condenser coils. If that air flow is blocked, energy performance will drop (see **INSTALLATION** above).

Check Power-Saver Switch

Many refrigerators have small heaters built into the walls to prevent moisture from condensing on the outer surface—as if the refrigerator doesn't have to work hard enough already! On some newer units, this feature can be turned off with an energy-saver or power-saver switch. Unless you have noticeable condensation, keep this switch on the energy-saving setting.

Defrost as Necessary

Manual defrost and partial automatic defrost refrigerators and freezers should be defrosted on a regular basis. The buildup of ice on the coils inside the unit means that the compressor has to run longer to maintain cold temperatures, wasting energy. If you live in a very hot, humid climate and don't use air conditioning, defrosting may be required quite frequently with a manual defrost model. After defrosting, you might be able to adjust the thermostat to a warmer setting, further saving energy.

RECOMMENDATIONS

1. Avoid putting hot foods directly in the refrigerator or freezer. Let them cool in the room first.

2 Cover foods, especially liquids. Otherwise they will release moisture into the refrigerator compartment, increasing energy use by the refrigerator.

3. A full freezer will perform better than a nearly empty freezer. This can be especially true in the event of a power outage. If your freezer isn't full, fill plastic containers with water and freeze them.

4. If you have a freezer or second refrigerator that's nearly empty, turn it off. You'll do no harm to your refrigerator or freezer by turning it on and off periodically. If you won't be using it at all, unplug it and remove the door to be sure that children can't accidently get trapped inside.

5. Mark items in the freezer for quick identification so that you don't have to stand there with the door open.

CHAPTER 8
Cooking

COOKING APPLIANCES

Choosing cooking appliances is a lot more complex and confusing today than it was 20 years ago. Along with the standard range with four top burners and an oven or two, we now have down-vented ranges with pop-out grills, fancy cook-tops, separate ovens, microwave ovens, convection ovens, and a host of other smaller cooking appliances from slow-cook crockpots and bread ovens to high-tech toaster ovens.

Cook-tops

This discussion applies equally well to standard kitchen ranges and separate cook-tops, which are gaining in popularity. Ovens will be addressed separately.

Cook-tops are widely available for either electric or gas cooking. Gas burners are often preferred by people who like to cook because gas offers a greater level of control in the speed of cooking. All new gas ranges are now required to have electric ignition instead of energy-wasting pilot lights. A downside to gas cooking appliances is that gas combustion products are introduced into the house. Operate a ventilation fan that vents to the outside when using a gas cook-top or oven.

With electric cook-tops, there are a number of new types of burners on the market, some of which offer energy savings and greater cooking control. The most common electric burners in this country are exposed coils, but you can also buy models with solid disk elements, radiant elements under glass, or high-tech halogen or induction elements.

Solid Disk Elements. These have been very popular in Europe. They are more attractive than coils and they are easier to clean up, but they heat up even more slowly than electric coils. Furthermore, because solid disk elements take longer to heat up and because higher-wattage elements are generally used, energy consumption will be higher. Good contact between the pan and the burner is especially important with solid disk elements. Slightly rounded pots and pans will significantly reduce cooking performance and correspondingly increase energy usage.

Radiant Elements Under Ceramic Glass. Ceramic glass units offer excellent cleanability and they heat up faster than solid disk elements—though not as quickly as conventional coil elements. The energy efficiency of ceramic glass cook-tops is higher than coil or disk element cook-tops. As with solid disk elements, flat pans are important with this type of cooking surface.

Halogen Elements. Fairly new in this country, halogen elements use halogen lamps as the heat source under a glass surface. The lamp delivers instant heat and responds very quickly to changes in the temperature setting. However, the main mode of heating a pan with halogen lamps still comes from contact of the pan with the hot ceramic glass surface. Therefore, the heating efficiency of halogen units may not be better than with ceramic glass units, and halogen elements will provide only marginally faster speed. Improvements in heating speed will not generate enough energy savings to justify the higher cost of halogen cook-tops.

Induction Elements. This is the newest and most innovative type of cook-top to come along. The induction elements transfer electromagnetic energy directly to the pan where the heat is needed. As a result, they are very energy efficient—using less than half as much energy as standard electric coil elements. One catch is that they work only with ferrous metal cookware (cast iron, stainless steel, enameled iron, etc). Aluminum cookware will not work. When the pan is removed, there is almost no lingering heat on the cook-top. Currently, induction elements are available only with the highest-priced cook-tops. The extra cost is difficult to justify on the basis of energy savings alone.

Range Hoods and Downdraft Ventilation

Proper ventilation for cooking appliances is very important. The range hood should ventilate to the outside and not simply recirculate and filter the cooking fumes. This is especially important with gas ranges. But also be careful about the size of the fan—too large a fan can waste energy and possibly even pose health problems.

When a ventilation fan is operated, it depressurizes, or creates a slight vacuum in the house. To balance that pressure difference, cold air is sucked in through cracks in the walls and around windows (infiltration). That makes your heating system work harder, wasting energy. In some situations, that negative pressure can even prevent an oil or gas heating system from venting properly, causing back drafting

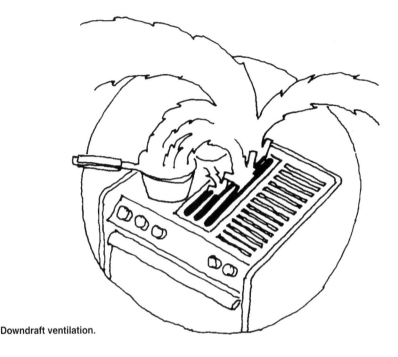

Downdraft ventilation.

of dangerous combustion gases into the house. This latter concern is especially serious with large downdraft ventilation fans used with some cook-tops and ranges. Ask about make-up air ducts for these ventilation systems. Some manufacturers, such as Jenn Air, are beginning to introduce such an option.

OVENS

In addition to the standard electric and gas oven, there are now convection ovens, microwave ovens, and combination models that work in one or more modes.

Conventional Ovens. With standard gas or electric ovens, self-cleaning models are more energy efficient because they have more insulation. But if you use the self-cleaning feature more than about once a month, you'll end up using more energy with the feature than you save from the extra insulation. If you're the type of cook who needs to peek into the oven all the time, buy a model with a window in the door, so you won't have to open the door all the time.

Convection Ovens. These are more energy efficient than standard ovens because the heated air is continuously circulated around the food being cooked. You get more even heat distribution, and temperatures and cooking time can be decreased. On average, you'll save about a third on energy use.

Microwave Ovens. Microwaves are very high-frequency radio waves. In these ovens, the energetic waves penetrate the food surface and heat up water molecules inside. Energy consumption and cooking times for certain foods are greatly reduced, especially small portions and leftovers to be rewarmed. Overall, energy use is reduced by about two-thirds. Because less heat is generated in the kitchen, you also save on air conditioning costs during the summer. Some microwave ovens include sophisticated features to further boost energy efficiency and cooking performance, such as temperature probes, controls to turn off the microwave when food is cooked, and variable power settings.

ENERGY-SAVING TIPS FOR COOKING

Whether or not you plan to buy a new range or other cooking appliances, you can probably save a lot of energy just by modifying your cooking habits. A few tips for energy-efficient cooking are listed below:

■ Full-size ovens are not very efficient when cooking small quantities of food. When cooking small- to medium-sized meals, it generally pays to use smaller microwave ovens, toaster ovens, or slow-cook crockpots (an insulated ceramic pot with an electric heating element). Several ways of cooking the same casserole are compared in the following table:

ENERGY COSTS OF VARIOUS METHODS OF COOKING

Appliance	Temp.	Time	Energy	Cost[1]
Electric oven	350°F	1 hr.	2.0 kWh	16¢
Convection oven (elec.)	325°	45 min.	1.39 kWh	11¢
Gas oven	350°	1 hr.	.112 therm	7¢
Frying pan	420°	1 hr.	.9 kWh	7¢
Toaster oven	425°	50 min.	.95 kWh	8¢
Crockpot	200°	7 hrs.	.7 kWh	6¢
Microwave oven	"High"	15 min.	.36 kWh	3¢

1. Assumes 8¢/kWh for electricity and 60¢/therm for gas.

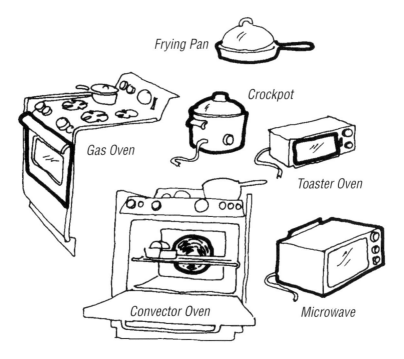

Frying Pan

Crockpot

Gas Oven

Toaster Oven

Convector Oven

Microwave

Deciding how to cook a meal is not as easy as it used to be. Some options result in significant energy savings.

■ If you have two ovens, use the smaller one whenever you can.

■ For soups and stews that require long cooking periods, using a crockpot will save a substantial amount of energy.

■ For stove-top cooking, consider using a pressure-cooker. By building up steam pressure, they cook at a higher temperature, reducing cooking time and energy use considerably.

■ Use the smallest pan necessary to do the job. Smaller pans require less energy.

■ With electric cook-tops, match the pan size to the element size. For example, a 6" pan on an 8" burner will waste over 40% of the heat produced by the burner.

■ Consider copper-bottom pans. These heat up faster than regular pans.

■ Keep the burner pans (the metal pans under the burners that catch grease) clean and shiny so they'll be more effective at reflecting heat up to the cookware. Blackened burner pans absorb a lot of heat, reducing burner efficiency.

■ With electric burners, solid disk elements, and radiant elements under ceramic glass, use flat-bottomed cookware that rests evenly on the burner surface. The ideal pan has a slightly concave bottom—when it heats up, the metal expands and the bottom flattens out. An electric element is significantly less efficient if the pan does not have good contact with the element.

■ With electric burners, you can turn off the burner just before the cooking is finished. The burner will continue radiating heat for a short while.

■ With gas burners, make sure you're getting a bluish flame. If the flame is yellow, the gas may not be burning efficiently. Have your gas company check it out.

■ To reduce cooking time, defrost frozen foods in the refrigerator before cooking.

■ With conventional ovens, keep preheat time to a minimum. Unless you're baking breads or pastries, you may not need to preheat the oven at all.

■ Try to avoid peeking into the oven a lot as you cook. Each time you open the door, a significant amount of heat escapes. Food takes longer to cook and you waste energy. Use your oven light and inspect through the window in the oven door instead.

■ Food cooks more quickly and more efficiently in ovens when air can circulate freely. Don't lay foil on the racks. If possible, stagger pans on upper and lower racks to improve air flow if you're baking more than one pan at a time.

■ Cook double portions when using your oven, and refrigerate or freeze half for another meal. It doesn't take as much energy to reheat the food as it does to cook it—not to mention the saved preparation time!

■ Use glass or ceramic pans in ovens. You can turn down the temperature about 25°F and cook foods just as quickly.

■ Use meat thermometers and timers to avoid overcooking. Overcooking not only spoils the taste and reduces nutritional value but also wastes energy.

■ If you have a self-cleaning oven, use the feature just after you've cooked a meal—that way, the oven will still be hot and the cleaning feature will require less energy. Try not to use the self-cleaning feature too often, and operate the ventilation fan when it's on.

■ With microwave ovens, keep the inside surface clean to allow more efficient microwave cooking. You can often cook foods right in their serving dishes, thus saving time and reducing the amount of hot water needed for dishwashing. (Always follow manufacturer's instruction on what type of cookware to use, which thermometers can be used, etc.)

CHAPTER 9
Dishwashing

AUTOMATIC DISHWASHERS

Most of the energy used by a dishwasher goes towards heating the water. In fact water heating accounts for approximately 80% of total energy use by dishwashers. Models that use less water therefore use less energy. Older dishwashers use between 8 and 14 gallons of water for a complete wash cycle. Dishwashers built since mid-1994 only use 7–10 gallons per cycle, thanks to national efficiency standards. Look at the manufacturer's literature for total water use with different cycles. If the literature does not provide this information, ask the salesperson.

Along with water use, a number of important features affect the energy use of dishwashers. These should be considered when shopping.

Booster Heater

Some dishwashers have built-in heaters to boost water temperature up to 140-145°F, the temperature recommended by manufacturers for optimum dishwashing performance. Some models will automatically

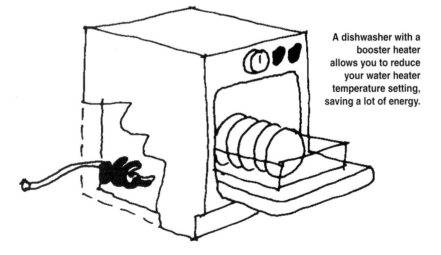

A dishwasher with a booster heater allows you to reduce your water heater temperature setting, saving a lot of energy.

boost water temperature if it is not hot enough, while other models re-quire preselecting the feature before starting the cycle. The advantage to the booster heater is that you can turn down your water heater ther-mostat, significantly reducing water heating costs. Each 10°F reduction in the water heater temperature setting cuts energy consumption from water heating by 3-5%. Check the dishwasher manufacturer's instruc-tions for the minimum recommended water heater temperature setting (usually 120°F).

Energy-Saving Wash Cycles

Most dishwashers have several different wash cycle selections. If a load of dishes is only lightly soiled, a "light wash" cycle will save energy by using less water and operating for a shorter period of time. A KitchenAid Imperial dishwasher, for example, uses 7.5 gallons for a "light/china" cycle, 10.75 gallons for a normal wash cycle, and 13 gal-lons for the "pots/pans" cycle. Energy-saving cycles can save $5 to $15 per year.

A number of new dishwashers models boast "dirt sensor" technol-ogy which automatically adjusts water use based on the turbidity of the wash water. Potential energy benefits from these models are hard to measure under the existing test procedure which requires testing on clean loads. Several reports on these dishwashers have reported un-even performance; it appears that on dirty loads these dishwashers consume the same amount or even more energy and water than less sophisticated models. Furthermore, some units did not adjust water levels down for the first two or three lightly-soiled loads washed after a heavily-soiled load. Until these issues are resolved, the additional pur-chase price of models with these automatic controls may not be justi-fied in terms of energy and water savings or performance.

Energy-Saving "No-Heat" Dry

An electric heating element is generally used to dry dishes at the end of the final rinse cycle. Most new dishwashers now offer an energy-saving no-heat drying feature. At the end of the rinse cycle, if the fea-ture is selected, room air is circulated through the dishwasher by fans, rather than using an electric heating element to bake the dishes dry.

BUYING A NEW DISHWASHER

EnergyGuide labels must be displayed on all new dishwashers sold. This label tells you the estimated energy use by the dishwasher.

However, you should understand some limitations to the En ratings for dishwashers. The rating is based on operating washer through 322 cycles annually on the normal setting. If you're considering a model with other setting options that you would use most of the time, your energy use could vary substantially.

Also, be aware that there are two dishwasher classifications: compact capacity and standard capacity. Compact models will use less energy, but they also hold fewer dishes. Compact models are usually 18 inches wide, compared to the 24 inch width of most standard capacity models. According to the U.S. Federal Trade Commission, to be considered "standard capacity," a dishwasher must be able to hold at least eight place settings of dishes. A compact dishwasher may actually result in more energy use if you have to run it more frequently.

Our list of most efficient dishwashers includes several models which are only 18 inches wide, yet they can accommodate eight full place settings of dishes, thanks to creative design of the dish racks. Since these models meet the FTC definition of standard capacity, they are included among the most efficient.

You should realize that manufacturers aren't required to test cleaning effectiveness on the same settings that they rate energy efficiency. Claims about performance may be based on a heavy-duty pot-scrubbing setting, while the energy ratings are based on the normal setting. Finally, some European machines may require frequent manual cleaning of the drain filters. In contrast, many domestic dishwashers use tiny grinders to dispose of food particles and keep filters clear automatically.

All of the dishwashers listed below qualify for the ENERGY STAR® label and are at least 30% more efficient than the 1994 standard.

MOST EFFICIENT DISHWASHERS

Brand	Model	Label Usage (kWh/yr)	Annual Energy Cost ($)
Asko	18*5	344	29
Asko	1375	377	31
Asko	1385	377	31
Asko	1585	377	31
Maytag	DWU9962	448	37
Frigidaire	MD*251R*	451	37
General Electric	GSS1800Z	451	37
Kenmore	253.14349	451	37
Kenmore	253.17345	451	37
Roper	RUD0800EB	451	37
Creda	DW*-34*	475	39
Crosley	LV-55*	475	39

MOST EFFICIENT DISHWASHERS (cont.)

Brand	Model	Label Usage (kWh/yr)	Annual Energy Cost ($)
Crosley	LVI-55*	475	39
Euroline	LV-55**	475	39
Fagor	LV-55US*	475	39
Fagor	LVI-55US*	475	39
Equator	*B-55	475	39
Equator	*I-55	475	39
Miele	G605	477	40
Miele	G805	477	40
Amana	DWA22A	518	43
Amana	DWA33A	518	43
Amana	DWA53A	518	43
Frigidaire	FDB345LF	518	43
Frigidaire	FDB4**RF	518	43
Frigidaire	MDB122RF	518	43
Gibson	GDB421RH	518	43
Bosch	SHI680*UC	524	43
Bosch	SHU53**UC	524	43
Bosch	SHU680*UC	524	43
Thermador	DW24*U*	524	43
Thermador	DWI246U*	524	43
Regency	330	526	44
Regency	660	526	44
Whirlpool	DU912PFG	531	44
Regency	990	534	44
Jenn-Air	DW1000	535	44

INSTALLING A DISHWASHER

When you install a dishwasher, try to position it away from the re-frigerator. The dishwasher produces heat and will increase energy consumption of your refrigerator. If it is a built-in dishwasher, you might be able to add extra insulation to the top, sides, and back when it's installed (check with the dealer first). This will both save energy and reduce noise levels. Use unfaced fiberglass batts.

USING A DISHWASHER FOR MAXIMUM ENERGY SAVINGS

Whether buying a new dishwasher or using an existing one, you may be able to save a considerable amount of energy by changing the way you operate it. Below are tips on how to save energy with dish washing.

■ Use energy-saving cycles whenever possible.

■ If your dishwasher has a booster heater, turn down your water heater thermostat. Most dishwasher booster heaters can raise the water temperature at least 20°F, so a setting of 120°F for your water heater should work fine. The washing cycle will take longer if the dishwasher has to boost the temperature, but unless you need to wash several loads in a row, this shouldn't be a problem.

■ Use the no-heat air-dry feature on your dishwasher if it has one. If you have an older dishwasher that doesn't include this feature, you can turn the dishwasher off after the final rinse cycle is completed and open the door to allow air drying. Using the no-heat dry feature or opening and air drying the dishes will increase the drying time, and it could lead to increased spotting, according to some in the industry. But try this method sometime to see how well it works with your machine.

■ Don't pre-rinse dishes before putting them in the dishwasher. Modern dishwashers do a superb job of cleaning even heavily soiled dishes. Scrape off food and empty liquids—the dishwasher will do the rest. If you must rinse dishes first, at least use cold water.

■ Wash only full loads. The dishwasher uses the same amount of water whether it's half-full or completely full. Putting dishes in the dishwasher

Depending on how you wash dishes by hand, you might actually save energy by switching to a dishwasher—if you use it wisely.

throughout the day and running it once in the evening will use less water and energy than washing the dishes by hand throughout the day.

If you currently wash dishes by hand and fill sinks or plastic tubs with water, it's pretty easy to figure out whether you would use less water with a dishwasher. Simply measure how much water it takes to fill the wash and rinse containers. If you wash dishes by hand two or three times a day, you might be surprised to find out how much water you're currently using. Whether or not you will save energy by switching from washing-by-hand to using a dishwasher depends on both the dishwasher and how you wash the dishes by hand.

■ Load dishes according to manufacturer's instructions. Completely fill the racks to optimize water and energy use, but allow proper water circulation for adequate cleaning.

CHAPTER 10
Laundry

WASHING MACHINES

Like dishwashers, most of the energy used by washing machines is for heating the water. Water heating accounts for about 90% of the energy use—even more than with dishwashers. The primary energy conservation strategies, therefore, involve using cooler wash and rinse cycles and reducing water use. Unlike dishwashers, washing machines can use cooler temperature water with perfectly adequate results for most clothes.

Horizontal-axis (usually front-loading) clothes washers are, in general, much more efficient than vertical-axis washers. Many horizontal-axis (H-axis) washers only use one-third as much water as vertical-axis machines, and thereby reduce energy use by two-thirds. Yet despite their water and energy savings, many studies show that H-axis washers clean clothes more thoroughly than conventional vertical-axis machines with agitators. Whirlpool has introduced the only vertical-axis model with efficiency comparable to horizontal-axis designs.

Horizontal-Axis Clothes Washers

The re-emergence of H-axis clothes washers on the American market is an exciting development for consumers interested in energy savings and environmental quality. In addition to attractive energy savings, the water savings from these machines is crucial in areas where water is scarce. (Although we sometimes use the terms interchangeably, "front-loading" and "horizontal-axis" are not necessarily synonymous: Staber Industries builds a top-loading H-axis machine.)

To understand how horizontal-axis washers use so much less water and energy, consider that in a conventional top-loader the tub must be filled with water so that all the clothes are kept wet. The agitator then swirls the water around to clean the clothes. In contrast, a front-loader needs less water because the tub itself rotates, making the clothes tumble into the water.

Front-loaders have always been popular in Europe, and in the past few years European manufacturers have increased marketing their products in the U.S. In response, Amana, Frigidaire, and Maytag introduced

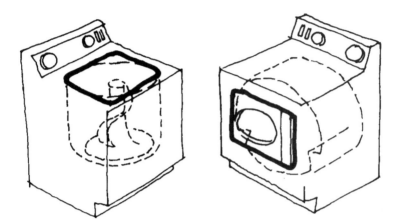

Horizontal-axis washing machines are far
more energy efficient than conventional top-loaders.

new H-axis machines in 1997, and other manufacturers may soon follow with their own designs. Many front-loaders permit stacking the
dryer on top of the washer, yet another benefit if space is tight. The
Maytag Neptune even features a tub angled up 15° for easier loading
and unloading.

At present, horizontal-axis clothes washers are more expensive to
purchase than vertical-axis washers; however, their substantial energy
and water savings translates into big money savings and a quick return
on your investment. Depending on your local energy and water rates
and the amount of laundry you do each year, you may realize annual
savings of $100 or more. If an H-axis washer cost $500 more to purchase than a conventional machine, your savings would be a tax-free
return on your investment of 20%.

A growing number of energy and water utilities around the country
recognize the benefits of efficient clothes washers, and are offering rebates to consumers who purchase qualifying machines. Call your energy and water utilities and ask if they provide rebates for high-efficiency clothes washers.

Efficiency

All new washing machines must display EnergyGuide labels to help
consumers compare energy efficiency. The EnergyGuide label for
clothes washers is based on estimated energy use for 416 loads of
laundry per year. But this value does not tell you the whole story for

washers because of variations in tub size and other fac
tional top-loading machines with smaller tubs may have
gyGuide ratings, but the smaller capacity may mean you   ⸱⸱ ʳun
the machine more often, so it may actually cost more to operate. However, although the tubs in horizontal-axis machines are often smaller
than top-loading tubs, their capacity for clothes is often the same because there is no agitator in the H-axis models and their tumbling tubs
can accommodate more clothes.

A better measure of a clothes washer's energy efficiency is given
by its energy factor, which is a combination of tub capacity and energy
consumption per cycle: $ft^3/(kWh/cycle)$. At present the energy factor is
not reported on EnergyGuide labels, but some manufacturers may
provide this information upon request. Use the list below to compare
different models.

MOST EFFICIENT CLOTHES WASHERS

Brand	Model	Capacity (cubic feet)	Energy Factor	kWh/cycle (elec. wtr. htr.)	kWh/yr (elec. wtr. htr.)	RMC (approx.)
Maytag	MAH4000	2.90	4.03	0.719	282	48
Frigidaire	FWTR445RF	2.65	4.01	0.661	259	<50
Frigidaire	FWT449GF	2.65	4.01	0.661	259	<50
Frigidaire	39012	2.65	4.01	0.661	259	<50
Frigidaire	39022	2.65	4.01	0.661	259	<50
Kenmore	29042	2.65	4.01	0.661	259	<50
Kenmore	29052	2.65	4.01	0.661	259	<50
General Electric	WSXH208V	2.65	4.01	0.661	259	<50
Gibson	GWT445RG	2.65	4.01	0.661	259	<50
Maytag	MHW2000	2.90	3.62	0.801	314	47
Maytag	ML*2000	2.90	3.62	0.801	314	47
Bosch	WFK2401UC	1.62	3.37	0.480	188	50
Miele	W1926	2.01	2.83	0.712	279	47
Equator	EZ 3600 CEE	1.90	2.75	0.681	267	31
Miele	W1903	1.69	2.66	0.635	249	39
Whirlpool (#)	LSW9245EQ	3.02	2.65	1.140	447	<50
Kenmore	29962	3.03	2.63	1.153	452	<50
Asko	11505(WM140)	1.60	2.62	0.615	241	49
Asko	13605(WM120)	1.59	2.62	0.615	241	42
Asko	20605(WM220)	1.60	2.62	0.615	241	39
Staber †	HXW2303	1.93	2.57	0.750	294	55
Asko	10505(WM90)	1.60	2.56	0.638	250	49
Miele	W1918	1.69	2.52	0.681	267	36
Miele	W1930	1.69	2.52	0.681	267	36
Creda	CWA 252	1.66	2.51	0.661	259	N/A
Creda	CWA 262	1.66	2.51	0.661	259	N/A

This Whirlpool model is the only vertical-axis washer available with efficiency levels comparable to
horizontal-axis designs.

† The Staber washer is a unique top-loading, horizontal-axis design.

Mechanical water extraction is much more efficient than thermal extraction (heating clothes in a dryer). After completing the rinse cycle, H-axis washers spin clothes faster than vertical-axis machines, so the remaining moisture content (RMC) of clothes is lower in an H-axis model. This means clothes need less time in the dryer, reducing wear-and-tear on the clothes and providing additional energy savings.

The Most Energy-Efficient Washing Machines

The most energy-efficient clothes washers are listed above, in order of efficiency based on each model's energy factor. For example, the Maytag MAH3000 uses more energy (0.800 kWh/cycle) than the Miele W1903 (0.620 kWh/cycle), but the Maytag's tub capacity (2.86 ft3) is so much larger than the Miele's (1.69 ft^3) that the Maytag's energy factor is higher and therefore the Maytag is considered more efficient: it can clean more clothes for a given amount of energy. Bear in mind that all of the machines listed here are at least twice as efficient—and some are nearly four times as efficient—as an average new top-loading, vertical-axis machine.

The last column in this table shows each model's estimated annual energy consumption; this is the value shown on EnergyGuide labels. We also include estimates of the remaining moisture content (RMC) after a typical cycle for each model: the lower the RMC, the less energy will be needed to dry the clothes. These energy savings from the clothes dryer are not included in either the energy factor or annual energy consumption (EnergyGuide) ratings, so you can consider them bonus savings, along with less wear-and-tear on your clothes.

All of these clothes washers qualify for the ENERGY STAR® distinction, and all of these models qualify for rebate incentives from those utilities offering rebates.

Wash and Rinse Cycle Options

Choose a washing machine that offers plenty of choices for energy-conserving wash and rinse cycles. The table below

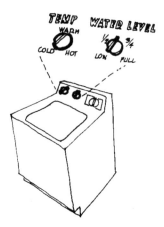

Look for energy-saving features on washing machines, such as cold-water rinse and multiple water-level settings.

shows the average energy use for conventional (vertical-axis) washing machines with both electric and gas-fired water heating (water temperature set at 140°F and 120°F). The dramatic differences in energy use with these different wash/rinse cycles are obvious. Warm wash cycles generally clean clothes perfectly well, and, with proper detergents and pre-soaking, cold water washing may be fine. Note: With oily stains, hot water may be required for satisfactory cleaning. You should experiment with the different cycle options and find one that meets your needs. Cold water rinses are just as effective as hot or warm rinses.

COST OF A LOAD OF LAUNDRY, VERTICAL-AXIS WASHER

	Electric water heater			Gas water heater		
Wash/rince settings	kWh used	Avg. cost per load (cents)[1]		Wash/rinse settings	Therms used	Avg. cost per load (cents)[2]
Water heater thermostat set at 140°F						
Hot/Hot	8.3	66		Hot/Hot	.329	20
Hot/Warm	6.3	50		Hot/Warm	.247	15
Hot/Cold	4.3	34		Hot/Cold	.164	10
Warm/Warm	4.3	34		Warm/Warm	.164	10
Warm/Cold	2.3	18		Warm/Cold	.082	5
Cold/Cold	0.4	3		Cold/Cold	—	3
Water heater thermostat set at 120°F						
Hot/Hot	6.5	52		Hot/Hot	.248	15
Hot/Warm	4.9	39		Hot/Warm	.186	10
Hot/Cold	4.3	27		Hot/Cold	.124	7
Warm/Warm	3.4	27		Warm/Warm	.124	7
Warm/Cold	1.9	15		Warm/Cold	.062	4
Cold/Cold	0.4	3		Cold/Cold	—	3

1. Assumes 8 cents per kWh.
2. Assumes 60 cents per therm.

Options to vary the cycle length have a very small effect on energy consumption, but some washing machines feature a "suds-saver" option that stores the wash water from one load of lightly soiled clothes to use for the next load. A presoak cycle may allow you to get the same cleaning performance using a warm water wash as you are used to getting with hot water—while using less total energy.

Water Level Controls

Most conventional washing machines use from 30 to 40 gallons of water for a complete (large) wash cycle. Choose a machine that lets you select lower water levels when you are doing smaller loads. For a given temperature cycle, energy use is almost directly proportional to hot water use. The lowest setting may use just half as much water as the highest. In general, you'll save energy by running one large load instead of two medium loads. Some models feature advanced electronic controls to adjust the water level automatically according to the size of the load.

Unfortunately, most manufacturers do not publish the actual water use of their machines in different settings, so it is difficult to compare one brand to another. Maytag machines use 20 gallons on the small capacity setting, 27 gallons on the medium setting, 34 gallons on the large setting, and 40 gallons on the extra-large setting.

Faster Spin Speed

Faster spin speeds can result in better water extraction and thus reduce energy required for drying. Mechanical water extraction by spinning is much more efficient than thermal extraction (heating clothes in a dryer). Front-loading washing machines generally spin at a faster speed than top-loaders.

DRYERS

Dryers work by heating and aerating clothes. Operation of both electric and gas-fired models is pretty straightforward. In terms of energy use, gas dryers are generally much less expensive to operate than electric models. Other than fuel type, the major energy consideration is whether the dryer senses dryness and turns off automatically and, if so, what the sensing mechanism is.

Automatic Shut-off

Dryers used to be controlled simply by setting a timer. You guessed how long it would take to dry a particular load of laundry and set the timer for 60 minutes or 75 minutes, etc. If the clothes came out dry, chances are you continued to use that setting. However, some types of clothes dry much more quickly than others. Overdrying can reduce the life of clothes. You can save a significant amount of energy by buying a model that senses dryness and automatically shuts off. Most of the better quality dryers today include this feature.

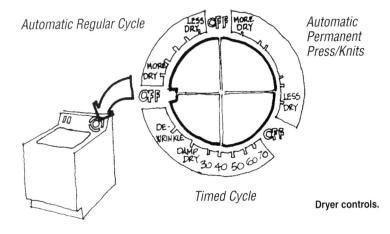

Automatic Regular Cycle

Automatic Permanent Press/Knits

Timed Cycle

Dryer controls.

The best dryers have moisture sensors in the drum for sensing dryness, while most only infer dryness by sensing the temperature of the exhaust air. The lower-cost, thermostat-controlled models may overdry some types of clothes, but even these are much better than timed-dry machines. Compared with timed drying, you can save about 10% with a temperature-sensing control, and 15% with a moisture-sensing control.

Electric Ignition on Gas Dryers

Electric ignition, rather than pilot lights, is now required for all new gas dryers.

Features That Reduce the Need for Ironing

While the energy used for ironing clothes is not technically part of the energy used for washing or drying clothes, it is very related. If clothes are taken out of the dryer while still slightly damp and then hung up, they may not need ironing. Moisture-sensing automatic shut-off helps here—especially if the dryer includes a feature to adjust the level of dryness desired. Wrinkling can also be reduced with a cool-down (fluff) cycle and by a feature that tumbles the clothes periodically after the end of the cycle if the clothes are not removed right away. Look for these features when shopping for dryers.

Dryer Exhaust Vent Hood

Along with your dryer, buy a dryer vent hood that blocks air infiltration. Dryer vent hoods are available that seal very tightly when the dryer

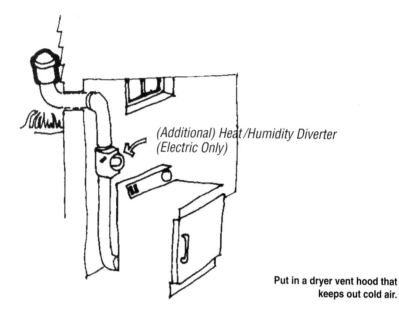

(Additional) Heat/Humidity Diverter (Electric Only)

Put in a dryer vent hood that keeps out cold air.

blower is not on. Standard dryer vent hoods have a simpler flapper that is not as effective. Ask your appliance salesperson about dryer vent systems and spend the extra $15 to $20 to buy a tight-sealing one.

INSTALLING WASHING MACHINES AND DRYERS

Try to install a washing machine as close to the water heater as reasonably possible and insulate the hot water pipes leading to it to minimize heat loss through the pipes. Also, if possible, locate the washer and dryer in a heated space. This is particularly important with dryers, which depend on heat to dry.

The most important part of the dryer installation is the exhaust system. In most cases—and always with gas dryers—the exhaust should vent to the outside, using as short and straight a section of smooth metal ducting as possible. Do not use flexible vinyl duct because it restricts air flow, can be crushed, and may not withstand high temperatures from the dryer. Use a vent hood on the outside that blocks air infiltration (see **Dryer Exhaust Vent Hood** above).

Electric dryers may be vented inside the home during the winter months if the house air is dry and if the air vent is properly filtered. Compact electric dryers in apartments and condos are sometimes installed for indoor venting. If the electric dryer is vented inside, watch for moisture on the windows, which would indicate that you're introducing too much moisture into the house.

LAUNDRY TIPS FOR MAXIMUM ENERGY SAVINGS

There are a number of easy ways to save energy with laundry, whether you're buying new appliances or not. Follow these suggestions whenever possible to keep energy use to a minimum.

■ Use lower temperature settings. Use warm or cold water for the wash cycle instead of hot (except for greasy stains), and only use cold for rinses. Experiment with different laundry detergents to find one that works well with cooler water. By presoaking heavily soiled clothes, a cooler wash temperature may be fine. The temperature of the rinse water does not affect cleaning, so always set the washing machine on cold water rinse.

■ Turn down the thermostat on your water heater. A setting of 120°F is adequate for most home needs. By reducing your hot water temperature, you will save energy with either hot or warm wash cycles.

■ Load the washing machine to capacity when possible. Most people tend to underload rather than overload their washers. Check your machine's load capacity in pounds, then weigh out a few loads of laundry to get a sense of how much laundry 10 or 18 or 20 pounds represents. Then use your eye to judge the volume of clothes for a load. Washing one large load will take less energy than washing two loads on a low or medium setting. Don't go to the other extreme and overload your machine, though. The clothes won't get clean and you may end up having to wash them again. When you don't have a full load, match the water level to the size of the load. Most washing machines, even older ones, offer several different settings. Note: Permanent press fabrics should be washed on a "permanent press cycle" to minimize wrinkling.

■ If washing lightly soiled clothes, use the suds-saving feature if it's available on your washing machine. This saves the wash water to be reused in the next load. Only use this feature, though, if the second load is to be washed right away.

■ When drying, separate your clothes and dry similar types of clothes together. Lightweight synthetics, for example, dry much more quickly than bath towels and natural fiber clothes.

■ Don't overdry clothes. Take clothes out while they are still slightly damp to reduce the need for ironing—another big energy user.

Overdrying also causes shrinkage, generates static electricity, and shortens fabric life. If your dryer has a setting for auto-dry, be sure to use it instead of the timer to avoid wasting energy.

■ Don't add wet items to a load that is already partially dried.

■ Dry two or more loads in a row, taking advantage of the heat still in the dryer from the first load.

■ Clean the dryer filter after each use. A clogged filter will restrict air flow and reduce dryer performance.

■ Dry full loads when possible, but be careful not to overfill the dryer. Drying small loads wastes energy. Overloading causes wrinkling and uneven drying. Air should be able to circulate freely around the drying clothes. If your washer and dryer are properly matched, a full washer load will be about the right size for the dryer as well.

■ Check the outside dryer exhaust vent. Make sure it is clean and that the flapper on the outside hood opens and closes freely. If the flapper stays open, cold air will blow into your house through the dryer and increase heating costs. Replace the outside dryer vent hood with one that seals tightly.

■ In good weather, consider hanging clothes outside and using totally free solar energy to do the drying.

CHAPTER 11
Lighting

Lighting accounts for 5-10% of total energy use in the average American home and costs $50 to $150 per year in electricity. That's not a huge amount, but it's enough to justify doing something about—especially when the advantages of energy-efficient alternatives are considered.

This chapter describes the most important new developments in energy-efficient lighting, includes listings of manufacturers and/or products, and provides recommendations on how to save energy with lighting around the home. One note on terminology: The lighting industry uses the term lamp to refer to the actual source of light—what the public usually calls the light bulb. In this chapter we use the terms lamp and bulb interchangeably.

TYPES OF LIGHTING

Incandescent Lighting

Most lighting in the home today is incandescent lighting. In an incandescent lamp, electric current heats up a metal filament in the light bulb, making it glow white-hot and give off light. The problem is that only 10% of that electricity is actually used to produce light—the rest ends up as heat. During the winter months, incandescent lighting is an expensive form of electric heat; during the summer months, it makes your air conditioner work harder.

Compact Fluorescent Lamps

The introduction of compact fluorescent lamps in the early 1980s revolutionized lighting. Compact fluorescent lamps use just one-quarter to one-third as much electricity as incandescent lamps, and they last up to ten times longer. They work in the same way as standard tube-fluorescent lamps, only the tube is smaller and folded over to decrease the amount of space required. The compact design allows them to be used in place of incandescent light bulbs. In fact, circular fluorescent

Compact fluorescent lights last, on average,
ten times as long as incandescent lights.

tubes ("circline" lamps) were an early application of fluorescents in incandescent fixtures. Today's compact fluorescent lamps continue this
evolution in technology.

The environmental benefits of these lights are dramatic. A single 20-
watt compact fluorescent lamp, for example, used in place of a 75-watt
incandescent light bulb, will save about 550 kWh over its lifetime. If
your electricity is produced from a coal-fired power plant, that savings
represents about 500 pounds of coal that would release 1300 pounds
of carbon dioxide and 20 pounds of sulfur dioxide.

Better yet, compact fluorescents are a profitable investment, saving
several times their purchase price through reduced electricity bills and
fewer replacement bulbs.

Integral vs. Modular Compact Fluorescent Lamps. All fluorescent
lamps need ballasts to operate. The ballast is a device that alters the
electric current flowing through the tube, which activates the gas inside, causing the tube to emit light. For this reason, compact fluorescent lamps are more complex to manufacture than standard incandescent light bulbs, and until recently they were a lot larger. Most compact
fluorescent lamps today have electronic ballasts that are integral with
the lamp. That is, the ballast and lamp are combined in a single unit,
which screws into a standard light bulb socket.

Because of the performance advantages of electronic ballasts, manufacturers of compact fluorescent lamps are switching their consumer product lines to integral units with electronic ballasts. Consumer acceptance of products with electronic ballasts is greater than those with magnetic ballasts.

Modular compact fluorescent lamps have separate ballasts and lamps. Currently, most of these use magnetic ballasts. Some modular ballasts, or adapters, screw into standard light bulb sockets. Others are hard-wired into light fixtures—i.e., the ballast is built into the fixture.

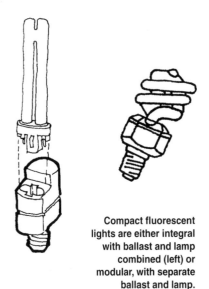

Compact fluorescent lights are either integral with ballast and lamp combined (left) or modular, with separate ballast and lamp.

The advantage of modular compact fluorescent lamps is that you don't have to replace the ballast when the fluorescent tube fails. Compact fluorescent lamps generally last about 10,000 hours, while magnetic ballasts are expected to last 50,000 hours or longer. Replacement lamps cost just a few dollars apiece, so in the long term you can save a lot with modular units.

Major manufacturers of compact fluorescent lamps are listed in Table 11.1.

Using Compact Fluorescent Lamps in Standard Fixtures. New products from all the major manufacturers are both smaller and brighter than older models. Rather than two folded-over fluorescent tubes, most of these new products have three, and a new generation of 4-tube lights is now on the market. By using more shorter tubes per lamp, the surface area of the lamp is increased allowing for more light from a smaller lamp. While they still aren't quite as small as standard incandescent light bulbs, these compact fluorescents now fit in most household fixtures designed for incandescent light bulbs. The smallest folded-tube compact fluorescent lamps are now just 4" to 5" long and 2.3" in diameter.

Despite the smaller size, even these compact fluorescent lamps—and many older ones—won't fit in all light fixtures. Sometimes they are too tall or the bases too wide. You may want to buy just one or

171

TABLE 11.1

COMPACT FLUORESCENT LAMPS

Duro-Test Lighting
9 Law Drive
Farifield, NJ 07004
800-289-3876
www.duro-test.com

GE Lighting
4400 Cox Road
Glen Allen, VA 23060
800-435-4448
www.ge.com/lighting

Lights of America
14 Tech Circle
Natick, MA 01760
800-876-0660
www.lightsofamerica.com

Osram Sylavania, Inc.
100 Endicott Street
Danvers, MA 01923
800-842-7010   978-777-1900
www.sylvania.com

Panasonic Lighting
1 Panasonic Way 4A-4
Secaucus, NJ 07094
201-348-5381
www.panasonic.com/MHCC/plmfeed.htm

Philips Lighting Company
200 Franklin Square Drive
P.O. Box 6800
Somerset, NJ 08875-6800
800-555-0050
www.lighting.philips.com

Real Goods
200 Clara Avenue
Ukiah, CA 95482
800-762-7325
www.realgoods.com

Sunpark Electronics Corp.
1815 West 205th Street, Suite 104
Torrance, CA 90501
888-478-6775   310-320-7880
www.sunpkco.com

two initially and try them in various fixtures around your house to check where they do and do not fit.

The latest development in compact fluorescent lamps is the sub-compact fluorescent lamp. These small, spiral-shaped lamps are significantly smaller than conventional compact fluorescent lamps, in fact several models are the same length as incandescent light bulbs. At this time, sub-compact fluorescents are available from only a few manufacturers and cannot be found in most retail channels. Table 11.1 includes contact information for manufacturers of sub-compact lamps.

Because they come in many shapes and sizes, it is likely that there is a compact fluorescent lamp that will fit your needs. Also, some compact fluorescents come as small, attractive globes, which can be installed in some ceiling sockets as attractive lamps by themselves, without any other fixture.

If a compact fluorescent lamp is too tall for your table lamp or standing lamp, you can buy a taller wire harp. If the compact fluorescent lamp is too wide, you can buy special adapters to spread the base of the harp. Be aware that compact fluorescent lamps can be somewhat heavier than standard light bulbs, although with electronic ballasts the extra weight is considerably less than with magnetic ballasts. The added weight could make some lightweight floor or table lamps unstable and easy to tip over.

For recessed downlights, spotlights, and track lights, some compact fluorescent lamps are too wide at the base to fit into the can or cone. Again, finding a product that fits may take some trial and error. Socket

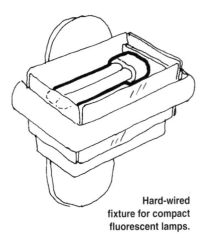

Hard-wired fixture for compact fluorescent lamps.

extenders are available and can help in some situations, although these may result in the tip of the lamp extending from the can or cone.

Never use compact fluorescent lamps in circuits that have dimmers unless the lights are specifically designed for that use. Dimmable products are beginning to enter the market in greater numbers. Compact fluorescents can be used in 3-way fixtures. They will operate on two of the three "on" settings, but only provide one light level.

ENERGY STAR®-labeled light fixtures suitable for use with compact fluorescent lamps are now available. A range of indoor and outdoor fixtures carry the ENERGY STAR® label. Many indoor models incorporate dimmers or two-way switches; outdoor models automatically shut off during the day or come equipped with motion sensors. Look for ENERGY STAR® for compact fluorescent fixtures which meet EPA's safety, reliability, and performance guidelines.

Economics. Compact fluorescent lights save a lot of money compared to incandescent lights. To figure out how, you have to look at both the purchase and operating costs to arrive at total life-cycle costs. It may surprise you to learn that the light bulbs you currently use cost a lot more to operate than they cost to buy. When you spend 75¢ for a 100-watt incandescent light bulb, for example, you're committing yourself to spending about $6.00 for electricity (at 8¢/kWh) over that bulb's expected 750-hour life.

With a high-quality compact fluorescent lamp, you might spend $10-20 to buy it (less if your utility company offers a rebate), but you save money in the long run, because it uses a lot less electricity and lasts a lot longer. Most compact fluorescent lamps last about 8,000-10,000 hours. The longer a particular light fixture is used each day, the faster a compact fluorescent replacement will pay for itself. This is shown in Table 11.2. If the light is on just two hours a day, it won't pay for itself until after the 3rd year. If it's on eight hours a day, though, you'll come out ahead during the first year. If modular compact fluorescent lamps are purchased, the savings will be even greater, because, as lamps fail, only the lamp itself and not the ballast has to be replaced.

Lower-cost compact fluorescents manufactured in China are now available in the U.S. While these products have historically been of questionable quality and limited life, recent joint ventures between major international manufacturers and Chinese lamp companies are leading to improved quality. The problem today is differentiating between the poor-quality and good-quality products. Before buying these

TABLE 11.2

SAVINGS ACHIEVED BY SWITCHING FROM INCANDESCENT TO COMPACT FLUORESCENT

Replace 75-watt incandescent with 20-watt integral compact fluorescent	Savings after 1st year	Savings after 2nd year	Savings after 3rd year	Savings after 5th year	Savings after 10th year
Lights on 2 hrs/day	($8.04)	($4.08)	($0.11)	$7.81	$27.62
Lights on 4 hrs/day	($4.08)	$3.85	$11.77	$27.62	$55.24
Lights on 8 hrs/day	$3.85	$19.70	$35.54	$55.24	$110.48
Lights on 12 hrs/day	$11.77	$35.54	$47.32	$82.86	$165.72

Assumptions:	75-watt incandescent	20-watt compact fluorescent
Lamp output (lumens)	1,220	1,200
Lamp life (hours)	750	10,000
Lamp cost ($)	0.75	12.00

Electricity cost: 8¢/kWh.
Numbers in parentheses are negative.

lamps, ask the salesperson about any complaints and their return policy. Chinese-produced lamps continue to improve and, as a result, low-cost compact fluorescent lamps may become an attractive option in the near future.

Tube Fluorescent Lighting

When you think of standard tube fluorescent lighting, what probably comes to mind is the buzzing, flickering bluish-white lights in supermarkets or offices that make colors look washed out and give some employees headaches. That is hardly the kind of light you want in your house. Well, times have changed. Tube fluorescent lighting has improved dramatically over the past ten years. Fluorescent lighting is available today that almost matches incandescent lighting in quality, accurately illuminating colors, and providing a pleasant warm-light atmosphere. Just as importantly, electronic ballasts that eliminate any noticeable hum or flicker are now available for tube fluorescent lighting. These newer tube fluorescent lights make a great deal of sense for indirect lighting around a room perimeter or above a bathroom mirror.

When shopping for tube fluorescent lighting fixtures and lamps, it helps to know what you want. Even at retail lighting stores, salespeople may not be familiar with some of the advanced products on the market. If you can't find what you want, go to a commercial lighting supplier.

For use in living areas, ask for electronic ballasts for tube fluorescent fixtures. These cost more than standard magnetic ballasts, but

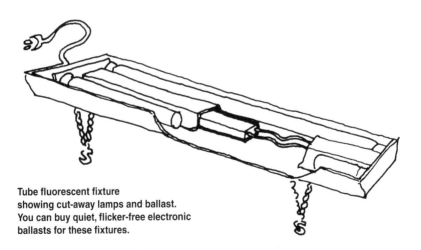

Tube fluorescent fixture
showing cut-away lamps and ballast.
You can buy quiet, flicker-free electronic
ballasts for these fixtures.

they are far more pleasant, and they save as much as 35% on energy use as well. Major manufacturers of electronic ballasts are listed in Table 11.3.

TABLE 11.3

ELECTRONIC BALLASTS FOR TUBE FLUORESCENT LAMPS

Advance Transformer Co.
800-372-3331
www.advancetransformer.com

Aromat Corp.
800-543-6204
www.aromat.com

L.G. Industrial Systems
888-544-7872
www.lgisusa.com

MagneTek
800-BALLAST
www.magnetek.com

Motorola Lighting
800-654-0089
www.motorola.com

Osram Sylvania
800-842-7010
www.sylvania.com

TABLE 11.4

HIGH-PERFORMANCE TUBE FLUORESCENT LAMPS

Manufacturer	Lamp (48")	Watts	CRI	Efficacy*
Osram Sylvania (800) 255-5042 www.sylvania.com	Octron (T-8)	32	75-85	90
General Electric 800-435-4448 www.ge.com/lighting	Trimline (T-8)	32	75-84	90
Litetronics Int'l. (800) 860-3392 www.litetronics.com	(T-8)	32	73-85	90-92
Philips Lighting Co. (800) 631-1259	Advantage-X (T-10)	40	80	93
Panasonic (201) 348-5381 www.panasonic.com/MHCC/plmfeed.htm	Tri-Color EX (T-10)	40	84	89

* Efficacy values are initial lumens/watt.

When selecting tube fluorescent lighting, don't be satisfied with standard cool-white or warm-white lamps. Look for products with high color rendition index (CRI). This is a measure of the ability of the light to illuminate colors accurately. Also, look for high efficiency (or efficacy, as it is called in the lighting industry). Lighting efficacy is measured in lumens (light output) per watt (electricity use). The best fluorescent lamps use special coatings ("trichromatic phosphors") to achieve both high CRI ratings and high efficacy. Some of the most energy-conserving fluorescent tube lamps are thinner in diameter and may require different fixtures and ballasts than standard fluorescents. Energy-saving, high-performance tube fluorescent lamps are listed by manufacturer in Table 11.4.

HID Lighting

High-intensity discharge or HID lighting is what you typically see along streets and in parking lots. HID lighting has advanced almost as quickly as fluorescent lighting in recent years. There are three types commonly used: mercury vapor, high-pressure sodium, and metal halide. Like fluorescent lamps, they require ballasts to operate, and most HID lamps take several minutes to warm up.

Mercury vapor lamps are still the most common for outdoor lighting around homes, but they are quickly becoming obsolete because of the higher efficacy of high-pressure sodium and metal halide lights. High-pressure sodium lamps are available with efficacies as high as 140 lumens per watt, though the light is somewhat yellowish. Metal halide lights produce a whiter light, closer to incandescent in quality, but the efficacy is lower than that of high-pressure sodium.

The primary place you will use HID lighting at home is outdoors: to light up the driveway, swimming pool, tennis court, etc. But a few manufacturers have introduced products designed for indoor lighting where very high lumens are needed, such as some commercial facilities.

Halogen Lighting

Halogen or tungsten-halogen lighting has improved somewhat over the past few years and remains the lighting option of choice where high light quality or precise light focusing is required. A halogen lamp is really a specialized type of incandescent lamp, often featuring a parabolic aluminized reflector (PAR) to improve light focus. Halogen lamps are slightly more energy efficient than standard incandescent lamps, but not as energy efficient as fluorescent lamps. In situations where light is needed on a precise area, halogen lights may be a more effective choice than fluorescent lights due to this tight focusing feature.

However, halogen torchiere floor lamps, which have become quite popular in recent years, are actually quite inefficient, since they consume 300-600 watts of electricity yet direct the light to the ceiling. Even though a smooth white ceiling can reflect some light, bathing a ceiling with light from a halogen torchiere wastes much of the light quality and tight focusing benefits of halogen lamps. Halogen torchieres are particularly poor choices in rooms with non-white or textured ceilings. In addition, these low-cost light fixtures pose a fire hazard due to the extremely hot temperatures produced by their high-wattage bulbs.

Halogen torchieres are an example of a low-price technology that proves to be costly over the long term. For example, a 300-watt torchiere used just two hours per day will consume 219 kWh per year at an average cost of $18 per year. Many discount stores sell halogen floor lamps for about $15, so this floor lamp can easily cost more to operate each year than it cost to purchase in the first place! Put another way, halogen torchieres use so much power that using one can cancel the savings you would gain by replacing six 75-watt incandescent light bulbs with 18-watt compact fluorescent lamps.

Fortunately, several companies have begun making energy-efficient torchieres with compact fluorescent lamps. These new, attractive products feature full-dimming or three-stage dimming (three light levels) and are much safer than halogen floor lamps while using only a fraction of the electricity. Most of these products carry the ENERGY STAR® label for easy identification. See Table 11.9 for contact information.

Solar walkway lights.

Solar-Powered Walkway and Patio Lights

Another idea in lighting is the use of solar energy to power outdoor lights. During the daytime, a photovoltaic (PV) panel generates electricity, which is stored in a battery. At night, that stored electricity is used to energize the light. Some models are turned on manually, while others are turned on automatically by light-sensing controls or activated by motion-sensing devices. Most of these walkway or security lights require no wiring or installation other than pushing the stake into the ground, or screwing the fixture to a garage wall.

Most of the widely marketed solar walkway lights do not put out a whole lot of light—don't expect to read the newspaper under a $50 walkway light—but they are very useful for lighting the path to your door so your guests can find their way. Larger solar lights are available that do provide a lot of light, but these can be quite expensive.

Solar-powered outdoor lights can be found in many hardware and department stores, or purchased through catalog retailers of alternative energy and stand-alone power equipment (see Table 11.5).

TABLE 11.5

CATALOG RETAILERS OFFERING PV-POWERED LIGHTING

Alternative Energy Engineering
P.O. Box 339
Redway, CA 95560
800-777-6609
www.alt-energy.com

Backwoods Solar Electric Systems
1395-in Rolling Thunder Ridge
Sandpoint, ID 83864
208-263-4290
www.backwoodssolar.com

Golden Genesis Company
4585 McIntyre Street
Golden, CO 88403
800-544-6466
www.goldengenesis.com

Jade Mountain Import-Export Co.
P.O. Box 4616
Boulder, CO 80306
800-442-1972
www.jademountain.com

Real Goods
200 Clara Avenue
Ukiah, CA 95482
800-762-7325
www.realgoods.com

Solar Webb, Inc.
136 East Santa Clara #6
Arcadia, CA 91006
888-786-9322
www.solarwebb.com

Sunelco
P.O. Box 787
Hamilton, MT 59840
800-338-6844
www.sunelco.com

Sunnyside Solar
RR 4, Box 808
Green River Road
Brattleboro, VT 05301
802-257-1482
www.sunnysidesolar.com

ʌtrols

...dition to saving energy by using more energy-efficient lamps, you can also save by having the lights on for a shorter period of time or at a lower output level. The simplest lighting control strategy, of course, is to turn lights off when you leave a room. Even if you are leaving for just a few minutes, it saves energy to turn the lights off. This applies to both incandescent and fluorescent lights.

With closets and cabinets, you can buy switches that turn the lights on when the door is opened and turn them off when the door is closed, thereby saving energy—as long as you remember to shut the door.

If you're the type of person who forgets to turn off lights when you leave a room, consider installing occupancy sensors that automatically turn lights off once a room is vacant. Most of these work by sensing heat or motion. While used primarily in commercial buildings, they can save energy in the home as well. Some models are available for as little as $20, while most are more expensive. However, inexpensive controls may be incompatible with electronic ballasts. Motion-sensing controls for outdoor security lighting both save energy and discourage potential intruders.

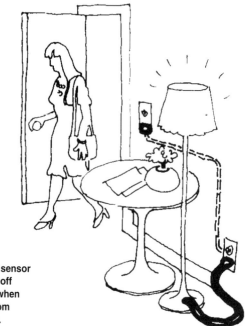

An occupancy sensor will turn lights off automatically when you leave a room to save energy.

TABLE 11.6

LIGHTING CONTROLS

Intermatic
777 Winn Road
Spring Grove, IL 60081
815-675-2321 ext.520
www.intermatic.com

Lightolier
631 Airport Road
Fall River, MA 02720
800-217-7722 508-679-8131
www.lightolier.com

UNENCO
185 Plains Road
Milford, CT 06460
800-245-9135
www.unenco.com

Watt Stopper
4101 East Park Blvd., Suite 138
Plano, TX 75074
800-879-8585
www.wattstopper.com

Novitas, Inc.
5875 Green Valley Circle
Culver City, CA 90230
310-568-9600
www.novitas.com

Light-sensing controls are increasingly being used to control outdoor lights. If you currently turn outdoor lights on at night, these controls are a great convenience, and they can save even more energy if connected to a timer to turn the lights off sometime late at night. A few of the manufacturers of the many lighting control products on the market are listed in Table 11.6.

Another control strategy with incandescent lights is to reduce light level and energy use with dimmers. Tube fluorescent lamps are dimmable only if they have dimming ballasts. Compact fluorescent lamps cannot be dimmed unless they are part of a complete dimming system, and dimming ballasts are not as efficient as nondimming ballasts.

TIPS FOR SAVING ENERGY AND MONEY WITH LIGHTING
Make Use of Natural Daylighting

Nothing's nicer than natural light, and in terms of energy use, nothing's more efficient. A single skylight or properly positioned window can provide as much light as dozens of light bulbs during the daylight

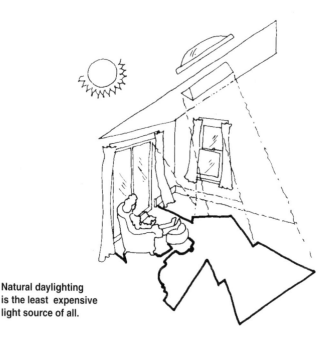

Natural daylighting is the least expensive light source of all.

hours. To benefit more from natural lighting you may need to rearrange your rooms somewhat—putting your favorite reading chair over by the south window, for example. Or you may want to go to more effort and install one or more skylights. To help get that light deeper into the room, you can paint your walls a light color and use reflective louvers or venetian blinds.

As you plan for natural lighting, don't forget that too much glass area on the east or west walls or on a south-facing roof can increase your air conditioning requirements. The best design balances passive solar heating, daylighting, and cooling considerations.

Reduce Background Light Levels and Rely More on Task Lighting

You can save a lot of energy by concentrating light just where it's needed and reducing background or ambient light levels. This strategy—called task lighting—is widely used in office buildings, but it makes just as much sense at home. Install track or recessed lights to illuminate your desk or the kitchen table where you do the crossword puzzle, and keep the ceiling lights off. Tungsten halogen lights can be focused precisely where you need the light.

Switch to Compact Fluorescent Lamps

Most of the lighting currently provided by incandescent lights can be provided just as well with compact fluorescent lamps. Replacing your incandescent lamps with compact fluorescents is the best way to save lighting energy in the average home. See the above discussion on compact fluorescent lighting and specific products that are available.

Where Design Permits, Use Tube Fluorescent Lighting

The best tube fluorescent lamps with new electronic ballasts are a far cry from what most of us think of as fluorescent lighting. They now make sense in places other than your garage or basement workshop. In

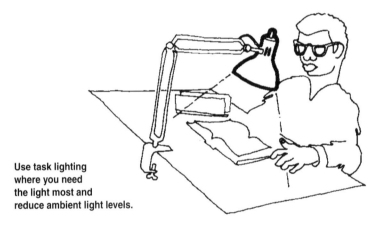

Use task lighting where you need the light most and reduce ambient light levels.

fact, they can provide very satisfactory (and energy-efficient) recessed lighting around the perimeter of a living room, or overhead lighting in kitchens and bathrooms. You can now even buy dimming fluorescent fixtures to vary ambient light levels precisely for excellent mood lighting.

Use Incandescent Lights Wisely

Higher-wattage incandescent light bulbs are more efficient than lower-wattage bulbs. It takes two 60-watt bulbs or four 40-watt bulbs to provide as much light as a single 100-watt bulb. In a fixture that holds several bulbs, you'll save by using a single higher-wattage bulb instead of several smaller bulbs (be sure to follow precautions on the fixture about maximum wattage, though).

"Watt Miser," "Supersaver," or "Econo-Watt" bulbs, available from the major lamp manufacturers, use 5-13% less energy than standard bulbs. They sometimes cost a little more, but that extra cost is more than made up for through energy savings. Halogen lights offer even greater savings, and they last longer. Long-life and rough-service bulbs, on the other hand, are less efficient and not recommended except in locations where changing bulbs is difficult, or where rough conditions cause standard bulbs to fail prematurely. Also avoid the "energy-saver" buttons that are widely promoted for saving energy and extending bulb life. They reduce energy use, but they reduce light output even more.

One 100-watt incandescent light bulb provides as much light as two 60-watt bulbs and uses less electricity.

Turn Lights Off When You Leave the Room, or Install Occupancy Sensors

Get in the habit of turning off lights when you leave a room. If you're forgetful, consider buying occupancy sensors that automatically turn lights off when you leave a room.

Install Energy-Saving Floodlights Outdoors

For outdoor lighting, use halogen, cold-start compact fluorescent, metal halide, or high-pressure sodium lights. Of these, halogen is the least energy efficient, and high-pressure sodium the most. If you generally leave outdoor lights on all night or (horrors) 24 hours a day, use light-sensing controls to turn the lights on at dusk and off at dawn.

Use Solar-Powered Accent Lights Outdoors

Solar walkway and patio lights are widely available in hardware and department stores or through catalogs. You can install them yourself in a few minutes without having to bury electric wires or hire an electrician.

Buying Energy-Efficient Lighting Equipment

Compact fluorescent lamps—and most of the other energy-efficient lighting products covered here—are still pretty new. They sometimes aren't available in great variety on the shelf with light bulbs at the supermarket or local convenience store, but most large home hardware centers now carry a wider array of efficient lights. Also, as more people learn about the benefits of efficient lighting, the products should become more readily available.

When you're looking for compact fluorescent lamps, ask for them in the stores where you normally buy light bulbs. If the stores carry them, buy them there. This will reinforce the company's decision to stock this new item. If they don't carry compact fluorescents, see if they will order them for you, or encourage them to consider stocking them from a wholesaler. But don't give up on buying compact fluorescent lights just because you can't find them locally.

There are quite a few mail-order and Internet suppliers of energy-efficient lighting products that carry everything you might need. If you can't buy locally, contact one of these companies listed in Table 11.7 and ask for a catalog of lighting products. You can also contact the manufacturers of compact fluorescent lighting products, or a wholesale distributor specializing in energy-saving lighting products, to find out what stores near you carry these products. A lighting wholesaler is listed in Table 11.8.

Last, but certainly not least, you should check with your local electric utility company to find out if they either sell compact fluorescent products or offer rebates on their purchase. Some utilities are subsidizing customer purchases of compact fluorescents lamps as part of demand-side management programs. The basic idea is that it is less expensive for utility companies to invest in energy conservation than to invest in building new power plants, so many utility companies are interested in helping you reduce your electric bills.

TABLE 11.7

MAIL-ORDER LIGHTING EQUIPMENT RETAILERS

Energy Federation, Inc. 14 Tech Circle Natick, MA 01760 800-876-0660 508-653-4299 www.efi.org	Jade Mountain Import-Export Co. P.O. Box 4616 Boulder, CO 80306 800-442-1972 www.jademountain.com
Harmony 360 Interlocken Blvd., Suite 300 Broomfield, CO 80021 800-869-3446	Real Goods 200 Clara Avenue Ukiah, CA 95482 800-762-7325 www.realgoods.com

TABLE 11.8

WHOLESALER OF ENERGY-CONSERVING LIGHTING PRODUCTS

Lighting Products
Fred Davis Corp.
93 West Street
Medfield, MA 02052
800-497-2970   508-359-3610

TABLE 11.9

ENERGY-EFFICIENT TORCHIERE MANUFACTURERS AND RETAILERS

Energy Federation, Inc.
14 Tech Circle
Natick, MA 01760
800-876-0660 508-653-4299
www.efi.org

LightSite
40 Washington Street, Suite 3000
Westborough, MA 01581-1012
800-379-4121
www.lightsite.net

Emess Lighting Company
1 Early Street
Ellwood City, PA 16117
800-688-2579   724-758-0707
www.emesslighting.com

CHAPTER 12
Other Energy Uses
In the Home

When we think about saving energy in the home, we generally focus on the obvious energy uses: heating, cooling, water heating, refrigerators, and so on—uses and products that the bulk of this book addresses. In many homes, though, there are miscellaneous uses of energy that are for the most part overlooked: such things as waterbed heaters, well-water pumps, pool filtering systems, engine block heaters, hot tubs, even aquariums for tropical fish.

While the energy use of each of these products may be relatively small on a national level, it can be quite large in individual houses. In fact, it is not too uncommon to find that one or more of these miscellaneous products can account for more energy use than your refrigerator, water heater, or even heating system. And when these miscellaneous products are looked at collectively on a nationwide level, the energy consumption is very significant: almost 20% of all electricity used in homes.

Let's take a look at these miscellaneous energy uses—both how they can affect energy use in an individual home and what the nationwide effect is. Table 12.1 lists the typical energy consumption of these household products and the saturation of these products nationwide. Table 12.2 looks at the nationwide energy consumption of these appliances.

One area of growing concern is the "leaking electricity" from home electronics and small household appliances that require direct current (DC). This equipment (TVs, VCRs, answering machines, electric-powered portable tools, etc.) draws power not only when they are in use, but also when the power is supposedly "off."

LEAKING ELECTRICITY

Home electronics and small household appliances that require direct current (e.g., TVs, VCRs, cordless phones, telephone answering machines, portable tools, rechargeable vacuums, etc.) draw energy not

TABLE 12.1

TYPICAL ENERGY CONSUMPTION AND PERCENT SATURATION OF MISCELLANEOUS ELECTRICITY USES IN THE HOME

Household product	Saturation	Typical energy consumption (kWh/yr)
Aquarium/terrarium	5-15%	200-1,000
Auto block heater	2-6%	150-800
Black & white television	20%	10-100
Bottled water dispenser	1-2%	200-400
Ceiling fan	40-60%	10-150
Clock	200%	17-50
Coffee maker	30-50%	20-300
Color television	200%	75-1,000
Computer	30%	25-400
Crankcase heater	5%	100-400
Dehumidifier	10-13%	200-1,000
Electric blanket	25-35%	75-200
Electric mower	5-8%	5-50
Furnace fan	50%	300-1,500
Garbage disposer	50%	20-50
Grow light & accessories	2-5%	200-1,500
Humidifier	8-15%	20-1,500
Instant hot water	1/2 -2%	100-400
Iron	20-40%	20-150
Pipe & gutter heater	2-5%	30-500
Pool pump	4-6%	500-4,000
Spa/hot tub (electric)	1-2%	1,500-4,000
Sumps/sewage pump	10-20%	20-200
Toaster/toaster oven	90-100%	25-120
VCR	80%	10-70
Ventilation fan	30-60%	2-70
Waterbed heater	12-20%	500-2,000
Well pump	10-15%	200-800
Whole-house fan	8-10%	20-500
Window fan	5-15%	5-100

Source: Leo Rainer (Davis Energy Group), Steve Greenberg & Alan Meier (Lawrence Berkeley National Laboratory), and U.S. Dept. of Energy surveys.

TABLE 12.2

ENERGY CONSUMPTION OF MISCELLANEOUS APPLIANCES IN THE UNITED STATES

Household product	Number in use (millions)	Energy use per unit (kWh/yr)	Total national consumption (TWh/yr)	Percent of total
Color television	200	250	50.0	5.15
Furnace fan	50	500	25.0	2.58
Waterbed heater	14	900	12.6	1.30
Pool pump	5	1500	7.5	0.77
Well pump	13	500	6.5	0.67
Video cassette recorder (VCR)	80	80	6.4	0.66
Aquarium/terrarium	10	600	6.0	0.62
Computer	40	130	5.2	0.54
DC power wall-pack transformers	200	25	5.0	0.52
Spa/hot tub	2	2300	4.6	0.47
Clock	180	25	4.5	0.46
Dehumidifier	11	400	4.4	0.45
Toaster/toaster-oven	86	45	3.9	0.40
Electric blanket	27	120	3.2	0.33
Cable TV boxes	40	80	3.2	0.33
Grow light & accessories	3	800	2.4	0.25
Ceiling fan	40	50	2.0	0.21
Coffee maker	40	50	2.0	0.21
Iron	32	50	1.6	0.16
Crankcase heater	5	250	1.3	0.13
Humidifier	11	100	1.1	0.11
Black & white television	20	40	0.8	0.08
Ventilation fan	54	15	0.8	0.08
Whole-house fan	8	80	0.6	0.07
Bottled water dispenser	2	300	0.6	0.06
Sumps/sewage pump	13	40	0.5	0.05
Garbage disposal	50	10	0.5	0.05
Pipe & gutter heater	3	100	0.3	0.03
Window fan	9	20	0.2	0.02
Electric yard tools	10	10	0.1	0.01
Total miscellaneous electric use	971[1]	1,6782[2]	162.8	16.80
Total household electric use	971[1]	10,0002[2]	970.0	100.00

1. 97 million houses is the total number in the United States.

2. Average per house.

Source: Adapted by John Morrill from work by Leo Rainer (Davis Energy Group), Steve Greenberg & Alan Meier (Lawrence Berkeley National Laboratory), Jennifer Thorne & Margaret Suozzo (ACEEE), and data from U.S. Dept. of Energy surveys.

189

only when they are in use, but also when the power is apparently off. This phenomenon is known as "leaking electricity." The average U.S. household leaks 50 watts of power constantly, amounting to about 440 kWh per year. Nationwide, this leaking electricity costs consumers over $3 billion per year.

TVs, VCRs, and cable boxes are responsible for more than half of this leaking electricity. These appliances draw power when off to support features such as instant-on, remote control, channel memory, and LED clock displays. Digital Satellite Systems also leak power; these devices consume an average of 13 watts when not in use, and DSS subscriptions are growing rapidly.

Considering that more than half of all American households have two or more color televisions (and over five million households have four or more color TVs), it is clear that home entertainment systems are becoming a major household energy use.

Fortunately, TV, VCR, and audio equipment manufacturers are working with the U.S. EPA to develop home electronics with low standby losses. Look for the Energy Star® logo when purchasing a new TV, VCR, or home stereo. Simply through improved circuitry, Energy Star TVs and VCRs will leak much less electricity than conventional electronics.

In addition to entertainment equipment, DC transformers that power answering machines, alarm systems, cordless phones, rechargeable tools—even electric toothbrushes—also leak electricity.

These square, black "wall pack" boxes each draw 2-6 watts of power as long as they are plugged in, even if the appliance is fully charged. The fact these boxes are drawing power can be demonstrated simply by grasping a wall pack that has been plugged in for a while: it will be warm to the touch. That warmth is simply electrical energy wasted as heat.

Researchers are working with transformer manufacturers to improve the efficiency of DC transformers to reduce these standby losses in new devices. In the meantime, consumers should unplug wall packs for appliances when they are not often used, such as hand-held vacuums, portable TVs, and rechargeable tools.

REMEMBER TO BOOST ENERGY EFFICIENCY WHENEVER POSSIBLE IN YOUR HOME— IT'S EASY ON THE POCKETBOOK AND THE PLANET

As you consider energy-saving opportunities in the home, look over this list. If you have a lot of these or other miscellaneous energy users in the home, there may be opportunities for savings. Recommendations are listed below for reducing energy use of a few of these household products.

■ *Furnace fans.* Refer back to Chapter 4 on heating systems. If your furnace is improperly sized, or if the fan thermostat is improperly set, the fan may operate longer than it needs to. If you're getting a lot of cold air out of the warm-air registers after the furnace turns off, this may be the case. Along with making you uncomfortable, the fan is wasting energy. On the other hand, if the fan shuts off too soon, heat from the furnace will be wasted. Have a service technician check the fan thermostat setting if you're unsure.

■ *Well pumps.* Well pumps are very common in rural areas. The amount of energy they use is dependent on how deep the well is, the pump quality, and the pressure controls. If the pump seems to be switching on more often than it should, there may be a leak in the system somewhere, or the pressure switch may not be functioning properly. Have the system checked out. You can also save on well pumping electricity costs by reducing your water use (see discussion on water conservation in Chapter 6).

■ *Spas and hot tubs.* As you can see from the tables, spas and hot tubs can use a tremendous amount of energy. If you have one, keep it covered with a tight-fitting insulated cover when it's not in use. If installing a spa or hot tub, insulate it well around the sides and bottom.

Make sure you use an insulated cover on a hot tub or spa.

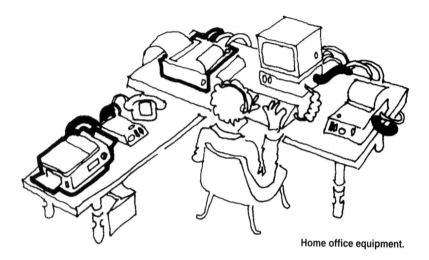

Home office equipment.

■ *Computers and home office equipment.* More and more people are working at home. As home office use increases, so does energy use by such equipment as computers, printers, copiers, and fax machines. Some of this equipment (especially color monitors and laser printers) consumes a great deal of energy. So look for the ENERGY STAR® label on any new computer equipment you purchase. The ENERGY STAR® label identifies efficient PCs, printers, faxes, and copiers. Fortunately, any increase in household energy use when you work at home is usually made up for by reduced energy use for transportation. In fact, the trend toward more home offices could have a significant impact on energy use for commuting.

■ *Waterbeds.* As many as 20% of homes have waterbeds, and most are heated with electric coils underneath. Waterbeds can be the largest electricity user in the home—exceeding even the refrigerator and water heater! Simple habits can significantly reduce energy waste from waterbeds. Regularly making the bed with a comforter can save more than 30%; insulating the sides of the bed can save over 10%. You might also consider putting the waterbed heater on a timer so that it doesn't waste energy throughout the day.

■ *Electric blankets.* Each electric blanket in a house uses an average of 150 kWh per year, according to various studies. If you use an electric blanket and frequently forget to turn it off in the morning, you can save energy by buying a simple timer control. Putting a second blanket or quilt over the electric blanket also saves energy, but be sure to check the electric blanket instruction for possible precautions against this practice. It is worth noting that electric blankets may actually save energy by allowing you to turn your thermostat down further at night.

■ *Block heaters.* If you live in a cold area and use a block heater to help your car start on cold mornings, you might be surprised at how much energy it draws. A 5-amp block heater could be using 14 kWh per day if you leave the car plugged in all the time. That's over a dollar per day! You will probably find that using the block heater for just a half-hour or so before you start the car will warm it up perfectly well. If you don't want to go out to the cold garage a half-hour before you leave, install an extra circuit with a timer-controlled switch in the house, and plug in the car when you get home. Then set the timer about a half-hour before you need to leave in the morning.

USE YOUR COMMON SENSE

Saving energy with many of these incidental energy-consuming products—and with other products throughout the house—is very easy. Most of the time, it just takes some common sense. If we all become a little more aware of the energy we use, we might just begin to solve some of our major environmental problems, and we'll end up with a little more money in our pockets as well.

APPENDIX 1
CO_2 Emissions and Energy Savings

The CO_2 emissions that result from using different types of fuel are listed in Table A.1. With this information, it's easy to calculate just how much CO_2 you are introducing into the atmosphere through your energy use. Simply look at your energy bills to find out how much fuel you are using: gallons of oil, therms of natural gas, kilowatt-hours of electricity, etc. Multiply that value by the quantity of CO_2 produced per unit of fuel in Table A.1.

TABLE A.1

CO_2 EMISSIONS FROM DIFFERENT ENERGY SOURCES

	CO_2 produced per unit of fuel	Lbs. of CO_2 produced per million Btus
Fuel Oil	26.4 lbs. CO_2/gallon	190
Natural Gas	12.1 lbs. CO_2/therm	118
Gasoline	23.8 lbs. CO_2/gallon	190
Coal (direct combustion)	2.48 tons CO_2/ton	210
Wood[1]	2.59 tons CO_2/cord	216
Electricity (from coal)	2.37 lbs. CO_2/kWh	694
Electricity (from oil)	2.14 lbs. CO_2/kWh	628
Electricity (from natural gas)	1.32 lbs. CO_2/kWh	388
Electricity (weighted national average including all generation)	1.54 lbs. CO_2/kWh	450

1. If the wood is harvested on a sustainable basis, there is no net CO_2 emission because the growing trees absorb more CO_2 than is released when burning the wood.

For example, if you use 1,300 gallons of fuel oil during a heating season, you are responsible for the release of 34,320 lbs. of CO_2 emissions. If you carry out energy conservation measures that reduce your oil use from 1,300 down to 1,000 gallons per year, you will have reduced your CO_2 emissions by almost 8,000 lbs. (300 gal. × 26.4 lbs.

CO_2/gal.). Similarly, if you cut your annual electricity use from 9,000 kWh to 6,000 kWh, you would be responsible for about 7,000 lbs. less CO_2 emissions, if your utility company burned coal to produce electricity. The CO_2 savings would not be as great if your utility company used a fuel other than coal.

If you use wood as a heating fuel and if the wood is grown on a sustained yield basis (i.e., the woodlot is left intact and younger trees replace those that are cut), there is no net contribution of CO_2 because the growing trees absorb CO_2 out of the atmosphere.

Appendix 2
Understanding EnergyGuide Labels

Federal law requires that EnergyGuide labels be placed on all new refrigerators, freezers, water heaters, dishwashers, clothes washers, room air conditioners, central air conditioners, heat pumps, and furnaces and boilers. These labels are bright yellow with black lettering. This section describes the labels, and provides guidance for using the information they contain.

WHAT INFORMATION IS ON THE LABEL?

Sample EnergyGuide labels are reproduced here for a refrigerator, a room air conditioner, a clothes washer, a water heater, a heat pump, and a furnace. For the first two sample labels, see the numbered lists for a brief description of key label elements.

HOW DO YOU USE THE LABEL?

The line scale in the middle of the label shows how that particular model compares in energy efficiency with other models on the market of comparable size and type. You will see a range of lowest to highest.

For refrigerators, freezers, water heaters, dishwashers, and clothes washers, the range shows annual energy consumption (in kWh/year for electricity or therms/year for gas). The most efficient models will have labels showing energy consumption (represented by the downward pointing triangle labeled "This Model Uses"), at or near the left-hand end of the range. It will be close to the words "Uses Least Energy."

For room air conditioners, central air conditioners, heat pumps, and furnaces and boilers, the range is not energy consumption, but rather the energy efficiency ratings for these products (EER, SEER, HSPF & SEER, and AFUE, respectively). Therefore, labels on the most efficient models will show "This Model's Efficiency" at or near the right-hand end of the range, close to the words "Most Efficient."

The label will not tell you who makes the more efficient models, or if they are available locally. To find out about the most efficient models, look at the listings in this book.

Based on standard U.S. Government tests

ENERGYGUIDE

1 — REFRIGERATOR-FREEZER WITH
AUTOMATIC DEFROST WITH SIDE-MOUNT
FREEZER WITHOUT THROUGH-THE-DOOR
ICE SERVICE

CAPACITY: 23.2 CUBIC FEET —— **2**

MODELS MS82354DRA, MSB2354DRW

Compare the Energy Use of this Refrigerator with Others Before You Buy.

This Model Uses

792 KWh/Year —— **3**

▼

4 —— Energy use (kWh/year) range of all similar models

Uses Least Energy	Uses Most Energy
750	848

kWh/year (kilowatt-hours per year) is a measure of energy (electricity) use. Your utility company uses it to compute your bill. Only models with 22.5 to 24.4 cubic feet and the above features are used in this scale.

Refrigerators using more energy cost more to operate.
This model's estimated yearly operating cost is:

$69 —— **5**

Based on a 1996 U.S. Government national average cost of 8.67¢ per kWh for electricity. Your actual operating cost will vary depending on your local utility rates and your use of the product.

Important: Removal of this label before consumer purchase is a violation of Federal law (42 U.S.C. 8302).   077563.096

1. Type of appliance.
2. Size, make and model number.
3. This model's annual energy consumption.
4. Scale indicating range of annual energy consumption for models similar in size and type.
5. This model's estimated yearly operating cost.

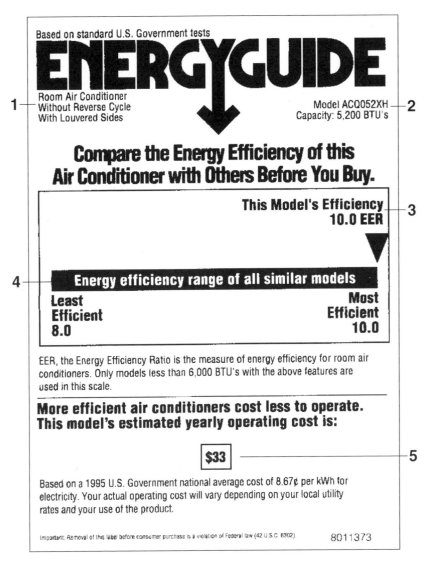

1. Type of appliance.
2. Size, make and model number.
3. This model's energy efficient rating.
4. Scale indicating range of efficiency ratings for models similar in size and type.
5. This model's estimated yearly operating cost.

A word of caution—the ranges shown on the labels are not updated frequently, and manufacturers are constantly introducing new appliances. For example, it is possible to find a model that is more efficient than the most efficient end of the range. Similarly, it is possible to find a model that is less efficient than the least efficient end of the range. On such a product, the label will include a statement explaining that the consumption or efficiency of that particular model was not available at the time the range was published.

The labels showing estimated annual energy consumption also show estimated annual operating costs, near the bottom of the label. This estimated cost is based on recent national average prices of electricity and/or natural gas, and assumes typical operating characteristics. For example, the energy use and cost estimates for dishwashers are based on six dishwasher loads per week; the estimates for clothes washers assume eight loads of laundry per week.

New furnaces and boilers must now carry EnergyGuide labels showing their annual fuel utilization efficiency (AFUE), the energy efficiency rating for these products. EnergyGuide labels on heating and cooling equipment refer customers to manufacturer's fact sheets available from the seller or installer. These fact sheets give further information about the efficiency and operating cost of the equipment under consideration.

EnergyGuide labels are not required on kitchen ranges, microwave ovens, clothes dryers, demand-type water heaters, portable space heaters, and lights. For these products, look for the energy-conserving features discussed throughout this book.

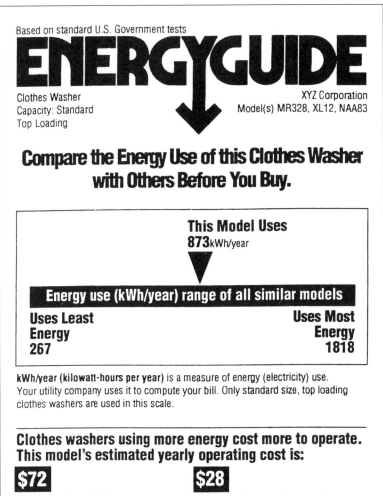

Based on standard U.S. Government tests

ENERGYGUIDE

Clothes Washer
Capacity: Standard
Top Loading

XYZ Corporation
Model(s) MR328, XL12, NAA83

Compare the Energy Use of this Clothes Washer with Others Before You Buy.

This Model Uses
873kWh/year

Energy use (kWh/year) range of all similar models

Uses Least
Energy
267

Uses Most
Energy
1818

kWh/year (kilowatt-hours per year) is a measure of energy (electricity) use.
Your utility company uses it to compute your bill. Only standard size, top loading
clothes washers are used in this scale.

Clothes washers using more energy cost more to operate.
This model's estimated yearly operating cost is:

$72
when used with an electric water heater

$28
when used with a natural gas water heater

Based on eight loads of clothes a week and a 1992 U.S. Government national average cost
of 8.25¢ per kWh for electricity and 58¢ per therm for natural gas. Your actual operating
cost will vary depending on your local utility rates and your use of the product.

important Removal of this label before consumer purchase is a violation of Federal law (42 U.S.C. 6302).

Label on an average clothes washer.
Note the cost comparison between electric
and gas water heating at the bottom of label.

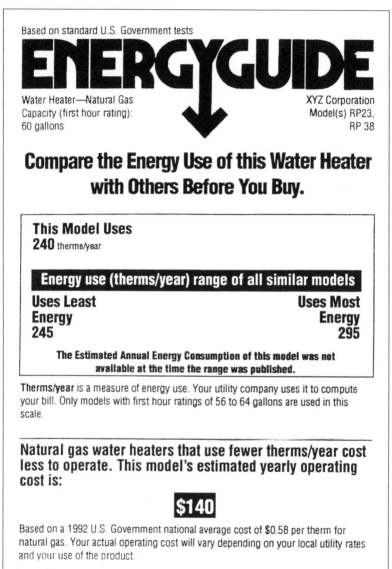

Based on standard U.S. Government tests

ENERGYGUIDE

Water Heater—Natural Gas
Capacity (first hour rating):
60 gallons

XYZ Corporation
Model(s) RP23,
RP 38

Compare the Energy Use of this Water Heater with Others Before You Buy.

This Model Uses
240 therms/year

Energy use (therms/year) range of all similar models

Uses Least
Energy
245

Uses Most
Energy
295

The Estimated Annual Energy Consumption of this model was not available at the time the range was published.

Therms/year is a measure of energy use. Your utility company uses it to compute your bill. Only models with first hour ratings of 56 to 64 gallons are used in this scale.

Natural gas water heaters that use fewer therms/year cost less to operate. This model's estimated yearly operating cost is:

$140

Based on a 1992 U.S. Government national average cost of $0.58 per therm for natural gas. Your actual operating cost will vary depending on your local utility rates and your use of the product.

Important: Removal of this label before consumer purchase is a violation of Federal law (42 U.S.C. 6302).

This highly efficient gas water heater's energy use is off the low end of the range, indicating it is more efficient than all models in prior years.

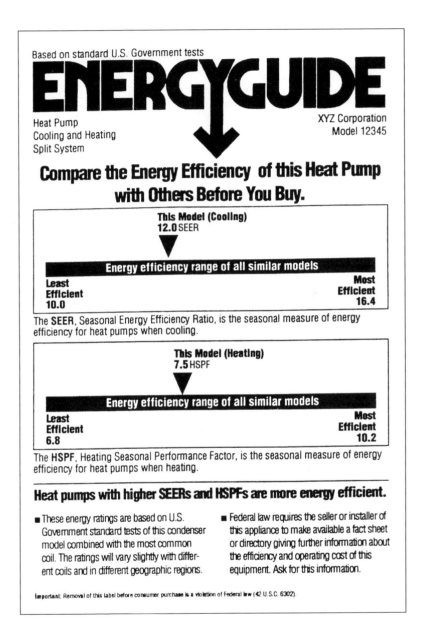

Based on standard U.S. Government tests

ENERGYGUIDE

Heat Pump
Cooling and Heating
Split System

XYZ Corporation
Model 12345

Compare the Energy Efficiency of this Heat Pump with Others Before You Buy.

This Model (Cooling)
12.0 SEER

▼

Energy efficiency range of all similar models

Least Efficient 10.0	Most Efficient 16.4

The **SEER**, Seasonal Energy Efficiency Ratio, is the seasonal measure of energy efficiency for heat pumps when cooling.

This Model (Heating)
7.5 HSPF

▼

Energy efficiency range of all similar models

Least Efficient 6.8	Most Efficient 10.2

The **HSPF**, Heating Seasonal Performance Factor, is the seasonal measure of energy efficiency for heat pumps when heating.

Heat pumps with higher SEERs and HSPFs are more energy efficient.

- These energy ratings are based on U.S. Government standard tests of this condenser model combined with the most common coil. The ratings will vary slightly with different coils and in different geographic regions.

- Federal law requires the seller or installer of this appliance to make available a fact sheet or directory giving further information about the efficiency and operating cost of this equipment. Ask for this information.

Important: Removal of this label before consumer purchase is a violation of Federal law (42 U.S.C. 6302).

Labels on heat pumps show both
cooling efficiency (SEER) and
heating efficiency (HSPF).

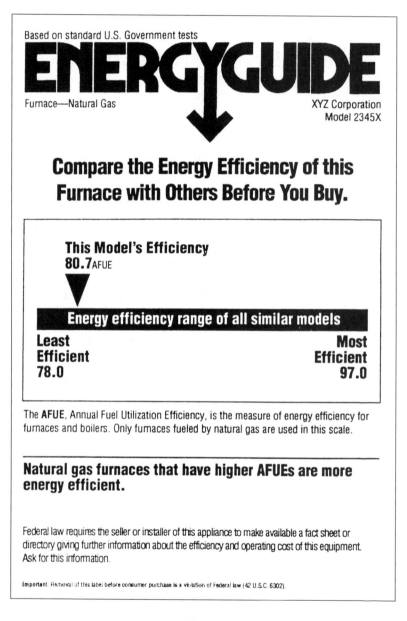

Based on standard U.S. Government tests

ENERGYGUIDE

Furnace—Natural Gas

XYZ Corporation
Model 2345X

Compare the Energy Efficiency of this Furnace with Others Before You Buy.

This Model's Efficiency
80.7AFUE

▼

Energy efficiency range of all similar models

Least	Most
Efficient	Efficient
78.0	97.0

The **AFUE**, Annual Fuel Utilization Efficiency, is the measure of energy efficiency for furnaces and boilers. Only furnaces fueled by natural gas are used in this scale.

Natural gas furnaces that have higher AFUEs are more energy efficient.

Federal law requires the seller or installer of this appliance to make available a fact sheet or directory giving further information about the efficiency and operating cost of this equipment. Ask for this information.

important: Removal of this label before consumer purchase is a violation of Federal law (42 U.S.C. 6302).

Furnace label shows annual fuel utilization efficiency (AFUE). This particular model is not very efficient.

APPENDIX 3
List of Manufacturers

This appendix lists phone numbers (and web sites, when available) for manufacturers and brand names of the high-efficiency equipment listed in this book. Toll-free numbers are listed when available.

We suggest you call the manufacturer if you cannot find a listed model in your local area. In most cases the manufacturer will refer you to its distributor in your area.

These numbers were compiled from the latest available information and verified for accuracy. However, we caution the reader that there has been quite a bit of consolidation among makers of space heating and water heating equipment in recent years, and the numbers here may change if mergers and acquisitions continue in these industries.

This list does not repeat the company phone numbers provided in the text for lighting products, instantaneous water heaters, and solar water heating systems; please refer back to those sections for complete information on those products.

Manufacturer	Phone Number	Web Site
A.O. Smith	800-845-1108	www.hotwater.com
Admiral	800-688-9920	www.maytag.com
Airpro	800-481-9738	www.york.com
AirQuest	800-458-6650	www.inter-city.com
Air Temp	800-283-4599	www.fedders.com
Amana	800-843-0304	www.amana.com
American Standard	800-900-9063	www.waterheating.com
American Wtr Htr	800-999-9515	www.americanwaterheater.com
Apollo Comfort	800-365-8170	www.stateind.com
Asko	972-238-0794	www.askousa.com
Axeman-Anderson	570-326-9114	
Bard	419-636-1194	www.bardhvac.com
Bock	608-257-2225	www.bockwaterheaters.com
Bosch	708-865-5200	www.boschappliances.com
Boyertown Furnace	610-369-1450	
Bradford-White	800-523-2931	www.bradfordwhite.com

Manufacturer	Phone Number	Web Site
Bryant	800-428-4326	www.bryant.com
Buderus	800-283-3787	www.buderus.net
Burnham	717-397-4701	www.burnham.com
Carrier	800-CARRIER	www.carrier.com
Central Environmental Systems	877-874-7378	www.york.com
Chambers	800-442-1230	
Clare Brothers	519-725-1854	
Climate Master	800-299-9747	www.climatemaster.com
Cold Point	888-832-4920	www.coldpoint.com
Coleman	877-874-7378	www.york.com
Columbia Boiler	610-323-2700	www.columbiaboiler.com
Comfort Aire	517-787-2100	www.heatcontroller.com
Craftmaster	800-999-9515	www.americanwaterheater.com
Creda	800-992-7332	www.creda.com
Crispaire/E-Tech	770-734-9696	www.crispaire-e-tech.com
Crispaire (GSHP)	800-841-7854	
Crosley	800-688-1120	www.crosley.com
Crown	215-535-8900	www.crownboiler.com
Daewoo	201-460-2501	www.dwe.daewoo.co.kr
Daikin	212-935-4890	www.acedaikin.com.sg
Danby	800-263-2629	www.danby.com
Day & Night	800-428-4326	www.bryant.com
DEC/Thermastor	800-533-7533	www.thermastor.com
DeMarco	512-335-1494	www.demarcoenergy.com
Ducane	800-489-6543	www.ducane.com
Dunkirk	716-366-5500	www.dunkirk.com
Emerson Quiet Kool	800-283-4599	www.fedders.com
Energy Kinetics	800-323-2066	www.energykinetics.com
Energy Utiliz. Systems	800-432-8387	www.euspoolansspaheaters.com
Equator	800-935-1565	www.equatorappl.com
Estate	800-253-1301	www.whirlpoolcorp.com
Evcon	800-481-9738	www.york.com
Fedders	800-283-4599	www.fedders.com
Florida Heat Pump	954-776-5471	www.fhp-mfg.com
Fraser-Johnston	877-874-7378	www.york.com
Friedrich	210-225-2000	www.friedrich.com
Frigidaire	800-451-7007	www.frigidaire.com
General Electric	800-626-2000	www.ge.com
Gibson	800-458-1445	

Manufacturer	Phone Number	Web Site
Glowcore	330-273-7770	www.glowcore.com
Goldstar	800-243-0000	www.lgeservice.com
Goodman	713-861-2500	www.goodmanmfg.com
GSW Water Heating	800-447-6575	www.gsw-wh.com
Guardian	800-481-9738	www.york.com
Hampton Bay	800-283-4599	www.fedders.com
Heat Controller	517-787-2100	www.heatcontroller.com
Heat Transfer Products	800-333-9651	
Heil	800-458-6650	www.heil-hvac.com
Hotpoint	800-626-2000	www.geappliances.com/usa/hotpoint
Hydro Delta	412-373-5800	
Hydrotherm	413-564-5515	www.mestek.com/hydrotherm.html
Inter-City	800-458-6650	www.inter-city.com
Int'l Comfort Products	800-458-6650	www.icpusa.com
Intertherm	800-422-4328	www.nordyne.com
Janitrol	713-861-2500	www.janitrol.com
Jenn-Air	800-688-1100	www.jennair.com
Kelvinator	800-323-7773	www.kelvinator-intl.com
Kenmore (Sears)	local Sears	www.sears.com
Kirkland	800-422-1230	www.kitchenaid.com
KitchenAid	800-422-1230	www.kitchenaid.com
Lechmere	local Montg. Ward	www.wards.com
Lennox	972-497-5000	www.lennox.com
Lochinvar	615-889-8900	www.lochinvar.com
Luxaire	877-874-7378	www.york.com
Magic Chef	800-688-1120	www.maytag.com
Mammoth	800-328-3321	www.mammoth-inc.com
Marathon	800-321-6718	www.marathonheaters.com
Maytag	800-688-9900	www.maytag.com
Melvin	207-645-4212	
Metzger	800-736-2378	www.metzgermachine.com
Miele	800-843-7231	www.mieleusa.com
Miller	800-422-4328	www.millerfurnace.com
Mitsubishi Electronics	770-613-5840	www.mitsubishi.com
Modern Maid	800-843-0304	www.amana.com
Montgomery Ward	312-467-2000	
Newmac	902-662-3840	www.newmacfurnaces.com
New Yorker	800-535-4679	
Nordyne	800-422-4328	www.nordyne.com
Olsen Technology	519-627-0791	www.olsenhvac.com

Manufacturer	Phone Number	Web Site
Oneida Royal	315-797-1310	www.uticaboilers.com
Panasonic	800-211-7262	www.panasonic.com
Payne	800-428-4326	
Pennco	814-723-8371	
Peerless Heater Co.	800-858-5844	www.peerless-heater.com
Quasar	800-211-7262	www.panasonic.com
RCA	800-366-1900	www.rca-electronics.com
Reliance Water Heater	800-365-8170	www.stateind.com
Rheem	501-648-4900	www.rheem.com
Richmond	800-432-8373	
Roper	800-253-1301	www.roperappliances.com
Ruud	501-648-4900	www.rheem.com
Sanyo	800-421-5013	www.sanyo.com
Sears	local Sears	www.sears.com
Sharp	800-237-4277	www.sharp-usa.com
Signature 2000	local Montg. Ward	www.wards.com
Slant/Fin	516-484-2600	www.slantfin.com
Smith Cast Iron	413-568-9571	www.smithboiler.com
Speed Queen	800-843-0304	www.speedqueen.com
Staber Industries	800-848-6200	www.staber.com
State Industries	800-365-5793	www.stateind.com
Tappan	800-537-5530	
Teledyne-Laars	603-335-6300	www.teledynelaars.com
Tempstar	305-592-3510	www.tempstar.com
Thermador	800-735-4328	www.thermador.com
Thermo-Dynamics	570-385-0731	www.thermo-dynamics.com
Thermo-Products	219-896-2133	www.thermopride.com
Trane	903-581-3200	www.trane.com
Trianco-Heatmaker	603-335-6300	www.teledynelaars.com
Ultimate	716-366-5500	www.dunkirk.com
Unitary Products Grp	877-874-7378	www.york.com
Unus	703-237-2100	
Utica Boilers	315-797-1310	www.uticaboilers.com
Vaughn	508-462-6684	www.vaugncorp.com
Victa Hytemp	716-322-8812	
Viessmann	401-732-0667	www.viessman.com
Waterfurnace	800-436-7283	www.waterfurnace.com
Weatherking	501-648-4900	www.weatherking.com
Weil-McLain	219-879-6561	www.weil-mclain.com
Whirlpool	800-253-1301	www.whirlpoolcorp.com

Manufacturer	Phone Number	Web Site
White-Westinghouse	800-245-0600	www.westinghouse.com
Williamson	800-736-2378	www.metzgermachine.com
York	877-874-7378	www.york.com
Yukon	612-571-5241	www.alphaamerican.com

APPENDIX 4
For More Information

If you have additional questions or concerns that have not been answered in this book, we encourage you to check out the resources listed below. Government agencies, consumer and environmental organizations, trade associations, and private businesses can provide a wealth of information to help answer questions about home energy conservation, renewable energy resources, energy suppliers, and related matters. In addition, a range of energy-efficient products and services can be purchased through on-line and print catalogs. Many more web resources can be located through the "Energy Efficiency-Related Web Sites" page on the ACEEE web site (http://aceee.org).

GOVERNMENT RESOURCES

The **U.S. Department of Energy (DOE)** offers information to the public through the **Energy Efficiency and Renewable Energy Clearinghouse (EREC)**. EREC has a toll-free number and professional staff to assist consumers with questions about home energy use, renewable energy, recycling, and related issues. Call 800-363-3732.

DOE also provides useful consumer information through the **Energy Efficiency and Renewable Energy Network** web site. The site includes information on getting a home energy audit, tips for saving energy throughout the home, guidance on using solar energy, suggestions for remodelers and apartment dwellers, and a buying guide for purchasing energy-efficient appliances. Visit the site at http://www.eren.doe.gov/buildings/homeowners.html.

The **Environmental Protection Agency (EPA)** and DOE maintain a web site with information about ENERGY STAR® programs and products. The site describes each ENERGY STAR® program, provides news and updates on products, and allows users to search for ENERGY STAR®-qualifying products by brand, type, size, or model number. Check out the site at http://www.energystar.gov. Information and questions about ENERGY STAR® products can also be obtained by calling the toll-free ENERGY STAR® hotline at 888-STAR-YES.

Lawrence Berkeley National Laboratory offers information on energy-efficient appliances, residential energy software, utility programs, home energy ratings, and financing options as well as numerous reports, case studies, and newsletters (http://eetd.lbl.gov). The lab also offers an on-line interactive tool, the Home Energy Advisor, to analyze your home for energy savings (http://hes.lbl.gov/HES).

We also suggest you contact your state energy office and/or county extension service. These local agencies may have information of particular interest to your location.

CONSUMER AND ENVIRONMENTAL ORGANIZATIONS

The **Alliance to Save Energy** is a coalition of prominent business, government, environment, and consumer leaders. Energy-saving tips and information for consumers can be found at http://www.ase.org/consumer. The site also provides information for educators and the media. Call 202-857-0666 for additional information.

The **Consumer Federation of America/Consumer Research Council** provides information on ways to save energy and money throughout the home and answers to common questions about purchasing energy-efficient products and services. Call the Council at 202-387-6121 or visit their web site at http://www.buyenergyefficient.org.

Consumers Union publishes the well-respected *Consumer Reports*, a valuable resource for information on a range of products, including home appliances and consumer electronics. Product reviews provide ratings of key features, performance, repair history, and pricing in addition to information on energy use and efficiency. A web-based version of the magazine, along with an archive of product evaluations and reports, can be accessed for a modest fee at http://www.consumerreports.org.

The **Natural Resources Defense Council (NRDC)** is a nonprofit organization dedicated to protecting the world's resources and ensuring a safe and healthy environment for all. Visit NRDC on-line at http://www.nrdc.org to learn more about the connection between energy use and the environment, clean power sources, air pollution, and much more.

TRADE ASSOCIATIONS

The **Association of Home Appliance Manufacturers (AHAM)** provides information on a wide range of appliances through their "Just for Consumers" web site. The site also offers advice from the experts on appliance purchasing, use, maintenance, and repair. Visit the site at http://www.aham.org/indexconsumer.htm.

The **Building Performance Contractors Association** provides an introduction to the home performance industry for retrofitting buildings for health, safety, comfort, and energy savings.

The site can also help locate qualified contractors in your area. See http://www.home-performance.org.

Utilities and manufacturers are also useful sources of information. See Appendix 3 for phone numbers and web sites of manufacturers referenced in this book. Contact your local utility for information relevant to your location.

ENERGY SERVICES AND ENERGY-EFFICIENT PRODUCTS

Energy.com is an on-line source of consumer information about energy consumption and generation. Consumers can purchase energy-related products and, for those in states with a deregulated utility industry, choose an energy supplier or join a buying group through the site. Visit http://www.energy.com.

ENERGYguide sells a growing list of high-quality products to help reduce your energy costs along with the information you need to guide your decisions. Many of the products bear the ENERGY STAR® label from the Environmental Protection Agency. Enter your zip code and the site will determine what, if any, utility rebates apply to your locale and apply the rebate directly to your purchase. See http://www.energyguide.com.

Energy Federation Incorporated (EFI) offers a wide variety of water and energy saving products in their on-line catalog. The site will apply any applicable utility rebates directly to your purchase. Visit their site at http://www.efi.org.

Iris Communications has an on-line catalog of books, videos, and software for sustainable, environmentally sound construction. Visit their site at http://www.oikso.com/catalog or call for a free, print catalog (800-346-0104).

LightSite.net is a web site created by Ecos Consulting. Formed by some of the leading experts in industry, government, and the environmental community, Ecos is dedicated to making lighting cleaner, safer, and more affordable. This site has information about halogen torchieres and their alternatives plus a complete on-line catalog of ENERGY STAR®-labeled torchieres. Visit their site at http://www.lightsite.net.

Real Goods has been a leader in the sustainable products market offering goods, services, and advice on saving energy and using renewable energy in the home for more than 20 years. Over 10,000 products ranging from air and water purification devices to clothing, lighting, batteries, and toys are available on their web site (http://www.realgoods.com). Call 800-762-7325 for more information or to order a copy of their print catalog.

APPENDIX 5
Choosing Heating and Air Conditioning Contractors

Choosing a good contractor to install a new furnace or central air conditioner can be as important as the equipment you choose, because proper installation and maintenance is needed for the equipment to operate safely, reliably, and at maximum efficiency. Here are some suggestions for selecting a contractor, adapted in part from *Contracting Business* magazine.

■ If you already know a reputable heating and air conditioning contractor, that is a good place to start. If you don't, friends and relatives in the area can often give you recommendations.

■ Do not give your business to a company offering to give you an estimate over the phone without ever looking at the job to be done.

■ A well-trained, up-to-date contractor will not try to discourage you from purchasing high-efficiency equipment. Less-qualified companies may not keep their employees current with the latest technology, and therefore they may discourage you from new and better designs.

■ A good estimator will do a survey of your home and base his or her proposal on a heat-load calculation (or cooling-load calculation for air conditioning).

■ Many furnaces and central air conditioners are not properly sized for the homes they serve, because of improper sizing years ago and/or energy efficiency improvements to the building since the old equipment was installed. Better contractors will not use your existing equipment to size your new heating or cooling system.

■ A good estimator should also ask about any heating or cooling problems you have had with your old equipment, and offer understandable explanations or solutions.

■ Using their heat-load calculations, good contractors should be able to estimate the annual operating costs (energy bills) for the equipment they are proposing for your home.

■ A good company will give you a written bid (or proposal) outlining the equipment to be installed, the work to be done, and the price, including labor costs.

■ We suggest you get estimates from multiple contractors, but try not to let the lowest price be the main reason for selecting a contractor. Better contractors may charge more, but they probably offer greater value. Be skeptical of extremely low bids; those contractors may not be including all routine services and customary warranties, or they may be trying to unload outdated or unreliable equipment.

■ Reliable contractors are professional. Their people are prompt and courteous. How a company treats you now reflects how they will treat you if there is a problem. They should have an office or shop facility, and they should not be ashamed to have you visit them. An office or shop is an indication that the company has been in business and intends to remain in business.

INDEX

About the Authors

ALEX WILSON is the editor and publisher of *Environmental Building News*, a highly regarded newsletter on environmentally sustainable design and construction, based in Brattleboro, Vermont. He has also written extensively on energy and building technology; his articles have appeared in such magazines as *Architecture, Journal of Light Construction, Progressive Architecture, Home Mechanix, Home*, and *Popular Science.*

JENNIFER THORNE is on the staff of the American Council for an Energy-Efficient Economy. She has authored a number of reports on energy use in buildings, appliances, and consumer products. She also has experience educating citizens on a range of environmental and consumer issues. She received a bachelor's degree from Trinity University and a master's degree from the Yale School of Forestry and Environmental Studies.

JOHN MORRILL is on the staff of the American Council for an Energy-Efficient Economy. He is coauthor (with Peter duPont) of *Residential Indoor Air Quality and Energy Efficiency*, also published by ACEEE, and he has written numerous articles for professional and consumer audiences. He received his bachelor's degree from Clark University and a master's degree from the University of Virginia.

"Besides up-to-date comparisons of the latest HVAC systems and appliances, the guide includes an assortment of energy-saving tips."
—*Fine Homebuilding*

"Handy little resource book...for anyone who wants to save household energy costs. Well-designed, compact, nicely illustrated with useful line drawings and diagrams. Don't be without it."
—*The Workbook*

"...exactly what longtime NESEA members would expect from technical writer and former NESEA executive director Alex Wilson. . .comprehensive, easy-to-read prose. . .clear, accurate, and well-illustrated (over 100 cartoon-style illustrations perfect for the intended audience...convey needed information clearly)."
—*Northeast Sun*

"Whatever your question is about home energy use, this book can almost surely answer it."
—*Co-op Currents*

". . .plenty of tips to help save more money. . ."
—*Popular Science*

"A comprehensive reference guide (in plain English, no less) for homeowners who've chosen to wade through the sometimes murky waters of buying green, and money-saving, home improvement products."
—*The Flint Journal*

". . .an unparalleled reference. . ."
—*SunWorld* (Int'l. Solar Energy Society)

"With many diagrams, this is a readable, affordable guide...buy it!"
—*Library Journal*

"Wait! Before you buy a water heater, furnace, air conditioner, fridge, freezer, stove, washer dryer, or lightbulb, read this book...a solid foundation of energy-saving advice."
—*Garbage*

"Must reading for homeowners in the market for new appliances...contains a wealth of information on how to make the appliances you own now work more efficiently. The advice here will also save you hundreds of dollars a year in energy costs."
—*Better Homes and Gardens*

"Don't furnish your home until you've consulted [the consumer guide]. It's a must-have resource."
—*Metropolitan Home*

"It tells you which appliances you can replace for the biggest energy savings. It lays it out in words anyone can comprehend in real sentences and nice, neat diagrams."
—Coastal Maine News

"There are a number of sources of information about the energy efficiency of furnaces, boilers, water heaters, and air conditioners...perhaps the most useful to the typical consumer is the [consumer guide] published by the American Council for an Energy-Efficient Economy."
—Kansas Country Living

"You can save a lot of legwork by consulting [this book]...chock-full of tips on energy considerations of each category of appliances, as well as energy efficient tips for home renovators."
—The Green Consumer Letter

". . .a clearly-written, well-illustrated guide to insulation, air sealing techniques, heating and cooling systems, and energy-efficient lighting and appliances."
—The Journal of Light Construction

"The latest edition is out—get yours while it's hot. This little guide is a must have for anyone in the market for new appliances. It's advised reading for anyone with an energy bill."
—Garbage

"This book could have easily been titled The Encyclopedia of Home Energy Savings. It's the most comprehensive resource to home energy savings that I've seen. Every homeowner and environmentally conscious (or utility paying) renter should have a copy."
—Green Living

ALSO AVAILABLE FROM ACEEE

NO-REGRETS REMODELING
Creating A Comfortable, Healthy Home That Saves Energy

In this reader-friendly book, the editors of *Home Energy* magazine show how new technologies and building practices can turn typical remodeling projects into opportunities for long-term benefits that add comfort and value to a home. Using over 100 detailed illustrations, the book describes how to avoid recurring problems including drafts, overheating, mold and mildew, peeling paint, rotting roofs, and indoor air pollution. Guides to selecting heating, cooling, and ventilation equipment, water heaters, insulation, lights, and windows demonstrate the advantages of integrating energy efficiency into any remodeling plan.

Written to educate and inspire homeowners, *No-Regrets Remodeling* can do the same for architects, contractors, and utility reps seeking a tool for innovative customer service programs in a deregulated market.

No-Regrets Remodeling. Published by Energy Auditor & Retrofitter, Inc.

ISBN 0-9639444-2-8
1997, soft cover, 222 pages, glossary, index, 8½" x 11", $24.95 (postpaid)

ALSO AVAILABLE FROM ACEEE

GUIDE TO ENERGY-EFFICIENT COMMERCIAL EQUIPMENT

Using this guide, buyers can specify and select the most energy-efficient commercial equipment for their needs. Purchasers and users of commercial equipment will find practical information on how to reduce building energy consumption, improve building systems performance, and increase worker comfort and productivity. Designed particularly for purchasing officials and facilities managers, the guide focuses on:

- lighting options
- heating, ventilating, and air-conditioning (HVAC)
- motors for commercial applications

In each area, the guide defines energy efficiency and other key equipment performance features. It presents criteria for selecting efficient equipment and lists the most efficient commercial equipment on the market. It also provides information about how to operate and maintain the equipment.

Guide to Energy-Efficient Commercial Equipment. Published by ACEEE in cooperation with New York State Energy Research and Development Authority (NYSERDA).

ISBN 0-918249-30-9
1997, soft cover, 145 pages, 8½"×11", $30.00 (postpaid)

ALSO AVAILABLE FROM ACEEE

GREEN GUIDE
TO CARS AND TRUCKS
Model Year 1999

This easy-to-use consumer guide ranks cars and trucks according to environmental friendliness. Consumers can compare cars, vans, pickups, and sport utility vehicles by environmental impact, including air pollution, global warming, and fuel efficiency. The guide contains current information on:

- a "Best of '99" section featuring the "greenest" models in each class

- how to buy the cleanest and most efficient vehicle that meets your needs

- "Green Scores" for all 1999 makes and models listed by class—compact, midsize, and large cars; vans; pickups; and sport utilities

- istings for electric and other alternative fuels as well as gasoline and diesel vehicles

- tips on keeping your vehicle running cleanly and efficiently

- information on the environmental impacts of motor vehicles including global warming and the health effects of vehicle pollution

Green Guide to Cars and Trucks. Published by the American Council for an Energy-Efficient Economy.

ISBN 0-918249-36-8
1999, soft cover, 113 pages, 6"×9", $13.95 (postpaid)
Green Guide to Cars and Trucks: Model Year 2000
will be available in February, 2000.